Building Pathology and Rehabilitation

Volume 15

This book series addresses the areas of building pathologies and rehabilitation of the constructed heritage, strategies, diagnostic and design methodologies, the appropriately of existing regulations for rehabilitation, energy efficiency, adaptive rehabilitation, rehabilitation technologies and analysis of case studies. The topics of Building Pathology and Rehabilitation include but are not limited to - hygrothermal behaviour - structural pathologies (e.g. stone, wood, mortar, concrete, etc…) - diagnostic techniques - costs of pathology - responsibilities, guarantees and insurance - analysis of case studies - construction code - rehabilitation technologies - architecture and rehabilitation project - materials and their suitability - building performance simulation and energy efficiency - durability and service life.

More information about this series at http://www.springer.com/series/10019

J. M. P. Q. Delgado
Editor

Case Studies in Building Constructions

Springer

Editor
J. M. P. Q. Delgado
CONSTRUCT-LFC, Department of Civil
Engineering, Faculty of Engineering
University of Porto
Porto, Portugal

ISSN 2194-9832 ISSN 2194-9840 (electronic)
Building Pathology and Rehabilitation
ISBN 978-3-030-55892-5 ISBN 978-3-030-55893-2 (eBook)
https://doi.org/10.1007/978-3-030-55893-2

This Springer imprint is published by the registered company Springer Nature Switzerland AG
The registered company address is: Gewerbestrasse 11, 6330 Cham, Switzerland

Preface

Building pathology is an all-inclusive method in analysing buildings, with emphasis on building defects and performance, in order to develop a better understanding of building failures and to develop appropriate remedial and management solutions. The evolution of the building degradations can be interpreted as the continuous reduction in performance over time, and it is well known that if the performance decreases below the functionality limits, the functional service life limit is reached. So, rapid and correct identification of potential hidden defects observed in building constructions is extremely useful because a late identification of the problems can lead to high rehabilitation costs and loss of building warranties.

The main purpose of this book, *Case Studies in Building Constructions*, is to provide a collection of recent research works, case studies and real-life experiences on building construction, to contribute to the systematization and dissemination of knowledge. It includes a set of new real-life experiences in the building rehabilitation, building diagnostics, service life prediction and life cycle, and hygrothermal behaviour. This book intends to bridge the gap between current approaches to the surveying of buildings and the detailed study of defect diagnosis, prognosis and remediation.

This book presents several case studies related to building pathologies and a detailed set of references and further reading. This book is divided into seven chapters that intend to be a resume of the current state of knowledge for benefit of professional colleagues, scientists, students, practitioners, lecturers and other interested parties to network. At the same time, these topics will be going to the encounter of a variety of scientific and engineering disciplines, such as civil and materials engineering.

Porto, Portugal

J. M. P. Q. Delgado

Contents

Prince of Wales Museum: Revitalization of European Painting Store

Prafulla Parlewar

Abstract The Chhatrapati Shivaji Maharaj Vastu Sangrahalaya (formerly, known as Prince of Wales Museum of Western India) at Mumbai (Bombay) is an world heritage site built in year 1914 by British architect George Wittet. It is famous for it's Indo-Saracenic architecture and construction technology. The European painting store have many famous painting from the British period. From many years, this storage area was deteriorated with passage of time, however, the revitalization provided new design through analysis of building structural system. Proposed design was developed to sustain the loading conditions of the shelving system and paintings. This chapter in beginning discusses the building architecture and structural system. Then, it discusses the old condition and revitalization proposed for the store. Thus, this research illustrates approach for analysis of the structural pathology of similar historic buildings.

Keywords Historic Buildings · Museum · Load Test

1 Introduction

The Chhatrapati Shivaji Maharaj Vastu Sangrahalaya (formerly Prince of Wales Museum of Western India) at Mumbai (Bombay) was designed by British Architect George Wittet. In the early Twentieth Century, it was decided to establish a museum to commemorate the visit of Prince of Wales to India. In a resolution, it was decided to in June 22, 1904 that the building should have a handsome and noble structure befitting the site selected, and in keeping with the best style of local architecture. Then, the government of Bombay handed over a landmark crescent site located at Mahatma Gandhi Road in Fort area of Bombay. A open competition was invited for the design of the museum. Through this open competition, George Wittet was commissioned the work to design the museum in 1909. During same period, he

P. Parlewar (✉)
City Development Corporation (P) Ltd., Mumbai, India
e-mail: citycorporationindia@gmail.com

J. M. P. Q. Delgado (ed.), *Case Studies in Building Constructions*, Building Pathology and Rehabilitation 15, https://doi.org/10.1007/978-3-030-55893-2_1

designed the famous Gateway of India. Also, he worked in collaboration with John Bess in construction of historic General Post building in Mumbai. In year 2018, the museum building was declared as world heritage site. The building is famous for Indo-Saracenic architecture and construction technology. It has four large size art storages. These are main store, European painting store, basement store and annex store. All stores contains valuable items collected from historic time. The museum in famous for permanent and temporary exhibition halls containing exhibits from pre-historic period till modern civilization.

The research here illustrates various questions on historic building. How to revitalize such historic buildings? Which method can be used for structural analysis of the these type of historic buildings? How load testing can be undertaken for these type of historic buildings? The concept for revitalization of European painting store included reuse of existing vitalities, structural stability, modern technology and efficient space planning. This included new imposed load of sliding art shelving system and mobile art compactors. This proposed load was expected to affect the complete historic building. So, a load test was conducted to study the flexural behavior of the stone masonry and Reinforce Cement Concrete (RCC) members. In this test no flexural failure was found in the structural members of the building. Based on the result of load test, the revitalization proposal was implemented in the building.

2 Building Profile

The building was completed in 1905 for the visit of Prince of Wales to India. Later, it was used as the military hospital in 1914. In year 1923, it was opened as a museum by Lady Lloyd, the wife of Governor Sir George Lloyd. The building is three stored constructed in hybrid construction technology with use of stone masonry and Reinforce Cement Concrete (RCC) (Fig. 1). It is designed in a symmetrical design with two wings on both side of the entry. The building design is integrated with the landscape design to provide large open space.

A large portico provides entry to the building which leads to a three storied lobby. This lobby is central to connectivity and circulation of the building with dome on the roof. Lobby contains a staircase to go to all the floor. This lobby acts like an central core of the building for circulation and spatial design.

3 Indo-Saracenic Architecture and Construction Technology

The Prince of Wales Museum is designed on Indo-Saracenic architecture and construction technology (Figs. 1 and 2). This architecture, also referred as Indo-Gothic was developed by British architects in late Nineteenth Century [1]. Particularly, the

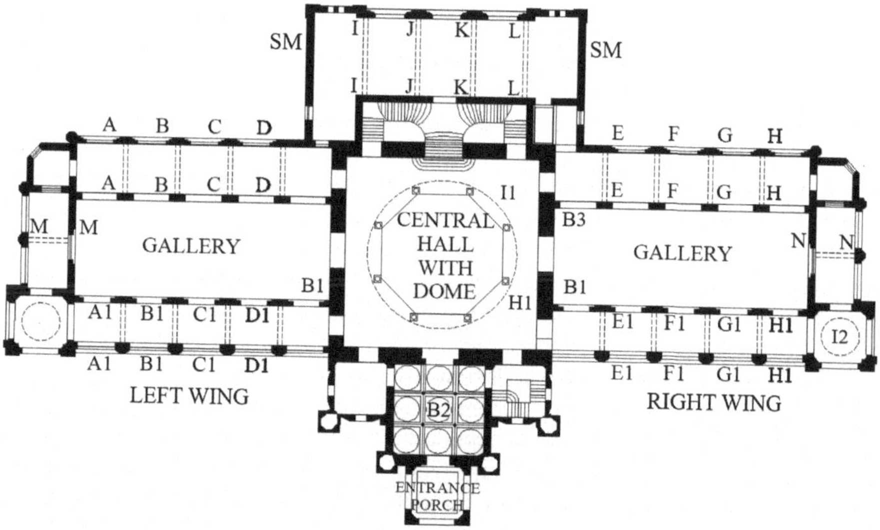

Fig. 1 Typical floor plan of building (I1—Islamic Large Span Dome, I2—Islamic Small Span Dome, I3—Islamic Turrets, B1—Gothic Corbelled Large Span Structure, B2—Coffered Dome System, B3—British Stone Arch Construction, H1—Hindu Decorative Column Supports, SM—Stone Masonary, AA to NN and A1A1 to H1H1—Reinforce Cement Concrete (RCC) Beams)

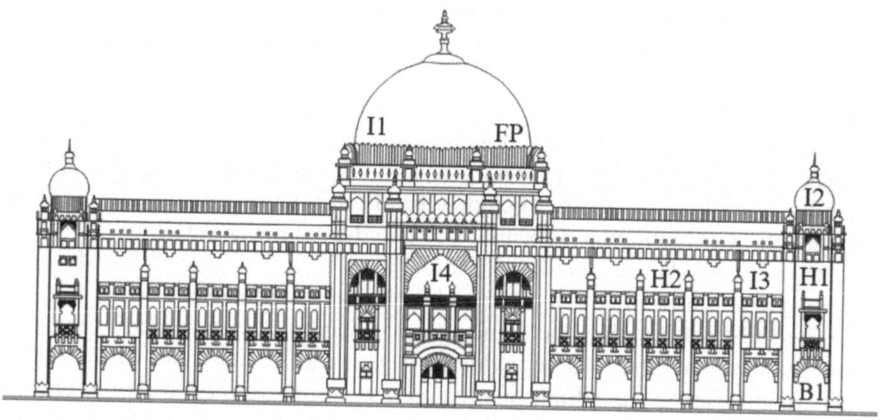

Fig. 2 Front elevation of building (I1—Islamic Large Span Dome, I2—Islamic Small Span Dome, I3—Islamic Turrets, I4—Islamic Pointed Arch, B1—British Stone Arch Construction, H1—Hindu Large Open Windows, H2—Hindu Corbelled Decorative Supports, FP—Floral Pattern in Dome)

movement began in 1870s which included mix of Indian architecture with colonial design. In the museum, architect have combined Indian, Islamic and British architecture and construction technology.

The overall design of building is based on the British colonial construction principals. It includes using grand scale which reflected the dominance of the British empire on the colony. Symmetry was an important principle of design with use of large central lobby. It has two wings to balance the three dimensional massing and two dimensional planning in the building. Architecture of building also incorporate the axial design system. Complete form of the building is designed on two axis like x and y. These axial pattern is followed in many small and medium design element to make a visually dominant, yet simple architecture for human vision.

Very interesting combination is designed by use of Indian architectural feature like railing, grills, and motifs. The central lobby of the building is designed with wooden railing and arches found in traditional Indian residential houses (Wada) in India. This design was famous during the medieval period in India. The lobby area is also consist of supporting columns similar to the Indian Temple architecture (Fig. 1). These provide structural support to the balcony of lobby. These are made form local stone with carvings by skilled artisan. The use of Indian architecture features are not just used in interior but also in exterior. The major architecture feature found is long cornice of exterior which provide support to the canopies. This is also made from stone with elaborate carvings from local craftsmen. These construction technology was prominent in Indian temple architecture.

The Islamic architecture and construction technology has been a major part of the design. The Islamic dome design on the main lobby provides a large spanning of spaces. The dome in central lobby is about 156 M in diameter. Significantly, a dome architecture is part of visual dominance in most of the Islamic architecture. Particularity, in Taj Mahal, it is found that dome is a central focus of design with a visibility from long distance. This dome has been found unique to its design to promote light and ventilation. It is also influenced from the architecture of various mosque constructed before Taj Mahal. Similar floral pattern is seen on the Mosques like Bibi Ka Makbara in Aurangabad and Gol Gumbaz in Bijapur, India (Fig. 2). The architect have strengthen the dome by use of Gothic design corbelling support in bottom. Moreocer, the Islamic architecture is found in many part of the building like pointed arches, handrail, pointed arch windows, turrets, grills, porch and others. Pointed arches are designed with the combination of Gothic stone masonry technology. Thus, there is combination of various construction technology found in Indo-Saracenic architecture.

4 European Painting Store

The European painting store is located on second floor of building which allows easy access to paintings from nearby galleries. Total 334 European paintings are present

Fig. 3 **a** European painting store, **b** lime plaster wall and **c** wall dampness

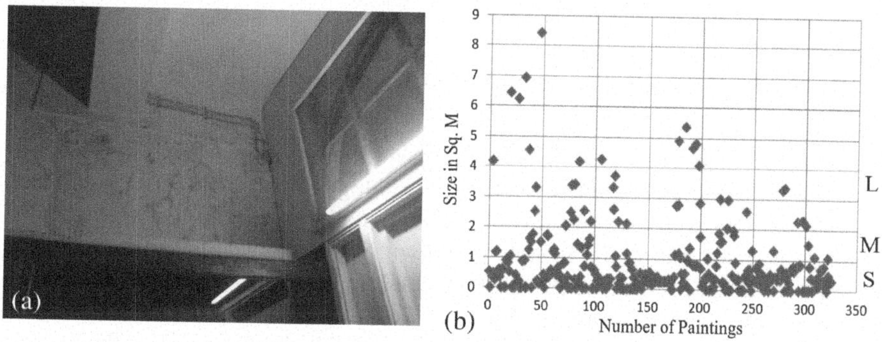

Fig. 4 **a** Dampness in building and **b** chart showing sizes of European paintings [S-Small (0–1.0 M), M-Medium (2.0–3.0 M) and L-Large (3.0 M and above)]

in the store. These painting range from the size 0.52 to 6.93 m^2 (Fig. 4b). Weight of these paintings ranges from 5 to 350 kg.

The physical condition of store dilapidated with passage of time because of no up-gradation from long time (Figs. 3 and 4). Some of the major damages were found in the walls, flooring, showcases, and ceiling. From long time, the windows and ventilators were not operated for ventilation. To developed micro-climate, large turbo fans were used to reduce the humidity and maintain the air changes. But, this was not sufficient to keep accurate micro-climate suitable for the paintings. Some of the paintings got affected because of lack of micro-climate. The store was also having insufficient natural and artificial lighting. Moreover, it required complete revamp of all building services.

The European painting store is an area of 20.75 m × 4.67 m with hexagon room at end (Fig. 5). The height of the store is 4.5 m. Outer walls of the store are made of load bearing construction in black stone. From many years, the paintings were inside the display racks. Some paintings were laying on the floor. During this period, there was no ventilation to the paintings. Due to lack of micro-climatic conditions, the acrylic paintings suffered deterioration. The main reason for deterioration was lack of air changes around the paintings.

All the wall of the European painting store are made of lime plaster. During the various renovation of store, the acrylic based wall paint was used on the wall surfaces. This type of wall paint reduces the breathing capacity of the lime plaster walls.

During the heavy rain, seepages were happening on the wall surface of the store. The main causes of these seepages were the damaged rainwater pipes located along the outer walls of the building. As these seepages were happening from long time, the lime plaster got damaged at many locations inside the store. The building design is based on innovative ventilation system with large windows and ventilators. These windows and ventilators were severely damaged inside the store.

5 Revitalization of European Painting Store

The concept for the revitalization of the European painting store was to reuse of existing vitalities, structural stability, modern technology and efficient space planning. The European painting were the most important vitalities of the store. Originally, the store was planned during the British period with a passive ventilation system, easy circulation and accessibility to the museum exhibitions. All these vitalities needed an innovative solution with accurate studies of the structural system of the building. In the initial studies, structural design was analysed for the building. Then, the loading system was designed as per the weight of existing paintings and proposed shelving system. A load test was conducted to understand the shear and flexural capacity of the structural members. After this test, structural system was found to be sufficient to carry the proposed load. Based on this, the proposal was executed for revitalization.

The original space planning was not accurate for the storage of painting and human circulation. This resulted into ineffectiveness to undertake the operation of removal, storage and conservation of the paintings. Proposed new planning, incorporated a office, painting conservation area, linear passage for circulation, sliding art shelving system, mobile art compactor, passive ventilation and physical improvements of the store. The proposed system of the mobile art compactors and sliding art shelving system vitalized the old painting (Fig. 5). This arrangement made it easy to organise historic paintings as per weight and size.

The ventilation for the painting is most important part of the any museum design. From many years, the European painting store was found to have inadequate ventilation which caused damages to various valuable historic European paintings. In a museum store, micro-climate conditions had to be design to avoid deterioration of expensive paintings. Generally, large changes in relative humidity and temperature cause damages to oil paintings, fabric and wooden frames. Ironically, the architect have already considered these aspect in the design of the Prince of Wales Museum. He planned the building floor with height of 7.25 m with ventilators on top and bottom of wall. Also, large windows were placed at equal position to make possible air changes. However, with passage of time these ventilators and windows were non functional and no air changes were happing in the store. This lead to the deterioration of the painting. The shelving system was designed for developing air changes between all the paintings.

The revitalization of European painting store had an comprehensive approach. It included replacing the existing lime plaster, restoration of flooring, electrical and furniture design. The deteriorated lime plaster in the wall was also affecting micro-climate in the storage area. Lime plaster allows moisture to easily circulate than a conventional cement plaster. Importantly, lime plaster is water repellent as well as breathable. It is flexible, anti fungal and provide strength to the building. This lime plaster inside the store was an important vitality to create a micro climate. Moreover, it was deteriorated at many places and causing dampness in the building. So, revitalization of the lime plaster provided micro-climate in the store. The lime plaster was made from composition of sand, water and slake lime. Specialized team of lime plaster worked on site to complete the lime plaster on walls.

The historic flooring is made of terrazzo tile of 1 ft × 1 ft size. This tile has been conserved during the revitalization process of the store. Efforts were done to replace the damaged tiles and clean with chemical for restoration. Lighting system was planned to increase the visibility of the painting during operation and conservation. Innovative ideas were implemented to make multi height doors for shifting the large scale paintings. Moreover, the specialised furnitures were designed like conservation tables, wall panels and officer cabin.

6 Design of Shelving System

The revitalization of European painting included re-arrangement of the paintings (Fig. 6). Studies on the existing paintings identified following aspects: (a) the sizes ranging from small to medium, (b) heavy weight of paintings, (c) framing system of the painting and (d) rear support system. About 334 numbers paintings are of size ranging between 0.52 and 6.93 m^2. We divided the paintings in three categories i.e small, medium and large. The small paintings range in size from 0.52 to 1.0 m^2. Whereas, the medium paining ranges in size from 1.0 to 2.0 m^2. The large painting

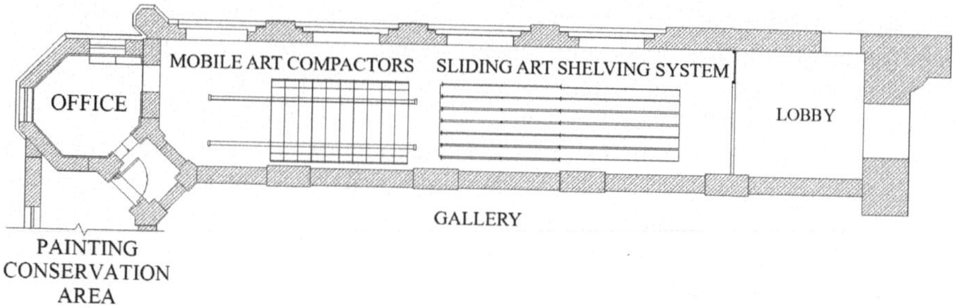

Fig. 5 Layout of European painting store

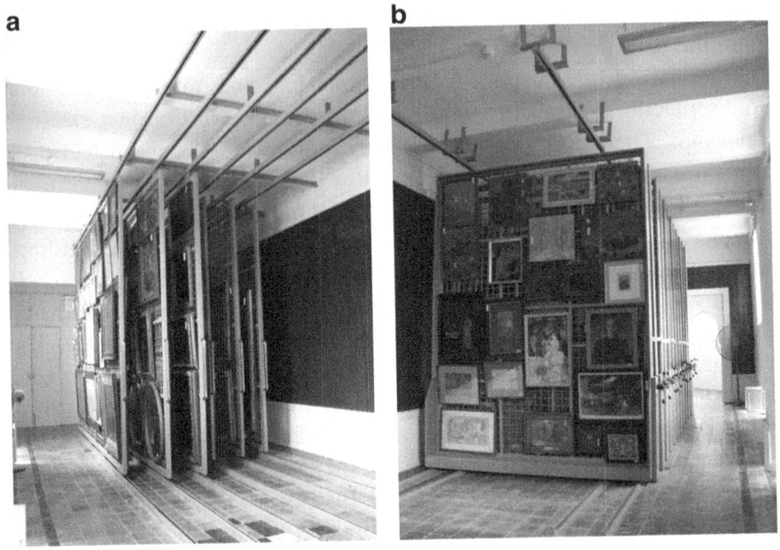

Fig. 6 **a** Sliding art shelving system and **b** mobile art compactors painting shelving system

ranges in size from 2.0 to 6.93 m^2. Some of the following painting are the largest in size paintings: (a) Words and Cloister (5.36 m^2), (b) Devid and Abigal (6.25 m^2), and (c) Landscape (6.93 m^2). Significantly, the weight of the painting was an important criteria for design of the shelving system to take load of paintings along with self load. Moreover, design was as per planned designed criteria of structural system of the heritage building. The weight of the largest painting is between 300 and 350 kgs.

The main purpose of shelving system was to organize the paintings in an order for easy functionality. Regularly, these paintings undergoes conservation, repairs and display. Furthermore, based on the requirements of display in the galleries, these paintings are shifted frequently to the exhibition halls. The revitalization plan developed two types of systems: (a) sliding art shelving system and (b) mobile art compactors painting shelving system. The sliding art shelving system included lateral movement system of size 5.09 m × 3.984 m. Each sliding art shelving system was designed from mild steel square pipe of size 75 mm as the main frame. This main frame was further divide into smaller squares of size 1.211 m × 0.929 m. Between these metal frames, metal mesh of 100 mm × 100 mm was fixed for keeping the paintings in vertical directions. Thus, the complete framing was supported with heavy duty six rollers in the four locations. These rollers were kept inside the channel section for easy movement. Finally, the system was fixed on the top beams and the floor. The design was developed in square pipes which distributes load in hierarchy (Fig. 7). This hierarchy of load distribution is as follows: (a) 10 mm thick mild steel rod mesh distributed the weight on the square pipe of size 1.211 mm × 0.929 mm (H1). (b) Then these square pipes distribute the complete load on the main frame of square pipes of size 5.09 mm × 3.984 mm (H2). (c) Finally, the main frame distribute

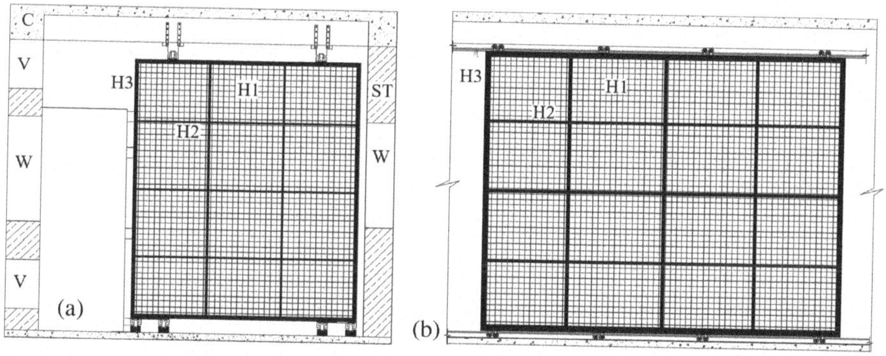

Fig. 7 **a** Design of mobile art compactors painting shelving system and **b** design of sliding art shelving system (C-Concrete, V-Ventilator, W-Window, ST-Stone Masonry, H1-Mesh, H2-Internal Load Member and H3-Main Frame)

the load on floor. This full load is disturbed on the load bearing system of the historic building (H3). Thus, the sliding art shelving system was designed to take load upto 350 kg. paintings. The painting of size 3.76 m × 2.24 m was easily install in vertical position on the sliding art shelving system.

All the historic paintings were fasten with U clipping system for hanging the paintings. This U clipping systems was carefully fixed under the expert supervision without causing damages to the historic paintings. Significantly, design offered easy installation and removal of the large paintings in the store areas.

Similarly, mobile art compactors were designed on hierarchy of the framing system for load distribution. These mobile art compactors were for the paintings of smaller size. These mobile art compactors were also designed to take load upto 350 kg. The loading system was distributed equally on the area of the store. The loading was equally distributed over the floor then transferred on the load bearing walls. The outer walls were further analysed to understand the loading of shelving system and paintings.

7 Structural Design Criteria

The building structural system is planned as load bearing system with three floors. In has typically two wings with central core. Thus, symmetrically distributing the load of the building. Four linear load bearing walls support the complete structure. Slabs are designed in Reinforces Cement Concrete (RCC) with beams resting on the load bearing walls. Such structure is also known as hybrid structure. Generally, hybrid structure is combination of load bearing system with Reinforce Cement Concrete (RCC) system.

Following are the silent points of the structural analysis for the revitalization proposal of European painting storage of the building: (1) As per the British Code

BS 8110, the load designed for the slab is 25.0 kN/mm^2 in museum buildings [2]. (2) As per the British Code BS 6399, the imposed load on the slab is 4.0 kN/mm^2 in museum building [3]. (3) The proposed imposed load on the museum floor is 1.5 kN/mm^2. (4) The load of single sliding art shelving system is 2100 kg and single mobile art compactor is 3750 kg. (5) During the load test no flexural failure were found in the structure.

8 Load Test

The strength evaluation of the concrete slab of the Prince of Wales Museum was undertaking by physical load testing method. As per the ACI—2008 Chap. 20 the physical load test is essential in case if there is doubt of the safety requirements of the structure. This test clarify the doubts about the shear strength. But, it can also be used for analysis of the problems of flexure capacity of concrete.

The load test was conducted for understanding the behaviour of the historic building for loading the proposed shelving system. Before the test, overall loading on the floor was analysed by estimating imposed load on the floor of the building. The total weight of the one sliding art shelving system was 350 kg. After fully loading, the paintings on all the sliding art shelving system, the total estimated load was 3990 kg. Weight of the mobile art compactors was estimated as 364 kg. The total weight of the fully loaded mobile art compactors is 6916 kg. Following methodology was followed for structural load test:

(1) Load Arrangement: The loading for the floor slab of the European painting was going to be affected the complete building. So, to understand the structural behaviour, load arrangement was planned on the floor slab of the European Painting store. The load was selected to maximize the deflection and stresses in the slab of the European painting store. To examine the structure, the load was uniformly distributed over the complete slab.

(2) Load Intensity: The total weight of the proposed sliding art shelving system and mobile art compactors was 5740 kg. The weight of paintings was estimated as 5166 kg. Based on these weights, total imposed load was found to be 1.50 kN/mm^2.

(3) Loading Criteria: The load test was conducted by imposing load intensity of 1.50 kN/mm^2 of sand bags. These bags were evenly placed on the floor to test the loading on the structure. This test continued for seven days. Structural members were checked frequently in these days to observe the failure of the structure. Following steps were followed for the loading criteria: (a) Initial diagnosis of the structural member was performed before starting the load test to understand magnitude, possible damages and need of any supports. A plan was prepared before conducting the test. (b) It was important to have sufficient safety measures to undertake the load test. These included the use of additional supports to the structural members during the test. (c) The failure criteria was established for the structural member. This criteria is a critical point at which the structure is expected to fail. (d) The records were conducted at 24 h interval for inspecting the deflections, rotational, strains or any

other effect on the structural member. (e) The imposed load of 1.5 kN/mm^2 become stabilised after some time. The observations were undertaken after 24 h for minimum seven days. (f) Increments were undertaken in the load upto certain intensity. After these increments, observations were also studies on the structural members. (g) The test was stopped after seven days. No affects to the structural member or any part of the building was found during the procedure. (h) The test load was removed and 48 h observations were conducted on the building. Photographic documentation was conducted of the every day. Finally, on the seventh day no failure was found in the structural members. The load test was successful which gave clearance to install the shelving system on the floor of the museum.

(4) Acceptance criteria: The following acceptance criteria were used for the load test. In the test no evidence of failure, spallings and crushing of compressed concrete was found in building. Thus, building was declared safe to undertake the proposed revitalization of European painting store.

9 Conclusions

The Prince of Wales museum have innovative structural system with Indo-Saracenic architecture and construction technology. The Indo-Saracenic architecture is generally combination of local architecture and construction technology. The building is designed with combination of Hindu, Gothic and Islamic architecture and construction technology. Design of building have combined these elements to developed aesthetic and innovative structural system. The building is composed of structural arrangement of Islamic system like domes, pointed arches and turrets. Also, building have Gothic construction technology. Exhibition areas are designed with corbelled design system of Gothic architecture. Furthermore, building is found to have combination of Hindu architecture elements like window, wooden arts and arches. Combination of these structural elements makes it an innovative building of Twentieth Century. The revitalization of European painting store in Prince of Wales museum included physical reorganization of spaces, design of new shelving system as per the structural criteria and improvement of the store area. Physical reorganization was undertaken by design of shelving system for the historic paintings. Design of these shelving system have sliding art shelving system and mobile art compactors designed as per the load criteria of the historic buildings. The design criteria included the possibility of loading the large size paintings. Furthermore, the load was disturbed such that the load is transferred on the load bearing walls. The load test concluded that there is no flexural failure of structural member. This test indicated that proposed structure is stable. As per the BS Code of 8110, the design criteria of the early Twentieth Century buildings were 25 kN/mm^2. The total imposed load on the floor of the European painting store was designed as 1.5 kN/mm^2. Thus, the design was found to be suitable for installation of the shelving system. Moreover, the revitalization of European painting store included the new design of office space, painting conservation area, lime plastering, water proofing, restoration of floor, lighting and furniture design.

The historic buildings of Twentieth Century are significant part of history but complex to understand the structural pathology through physical and numerical studies. Physical assessment of the building can be undertaken by understanding the building with geometric studies and structural testing. Historic buildings provides opportunities to understand the innovations in structural system of historic time. The geometric studies are very important to diagnose the structural system which include accurate documentation and photographic studies of the building. Generally, geometric studies are conducted by undertaking accurate measurements of structural members which indicates the overall configuration of structural system. Many different types of test can be conducted to understand the structural conditions and load imposition on the buildings. These are rebound hammer tests, ultrasonic pulse velocity tests, core sample test, cover test, reinforcement test, mortar test etc. Rebound hammer test and ultrasonic pulse velocity test are commonly used from last few years to diagnose the structural conditions of the historic buildings.

The Prince of Wales museum is an innovative structural system with Indo-Saracenic architecture and construction technology. Thus, the revitalization of European painting store had been successfully implemented in the historic building.

References

1. J. Sheeba, J.T.M. Dhas, A study of Indo-Saracenic architectural heritage. Int. J. Pure Appl. Math. **118**(22), 1737–1742 (2018)
2. British Standard Institute, *Structural Use of Concrete Part 1: Code of Practice for Design and Construction* (1997)
3. British Standard, *Loading for Buildings, Part 1, Code of Practice for Dead and Imposed Load*, BS6399: Part I, (1996)

Analysis of Temperature and Humidity Variation in the Stone Chamber of Takamatsuzuka Tumulus from Construction of the Conservation Facility to Before the Dismantlement of the Stone Chamber

Daisuke Ogura, Shuichi Hokoi, Takeshi Ishizaki, and Yonghui Li

Abstract Immediately after the discovery of the Takamatsuzuka Tumulus mural painting, which was discovered on March 21, 1972, in Asuka village in Nara Prefecture, Japan, various preservation measures were taken to protect the mural painting in its stone chamber, from local preservation to the dismantling of the stone chamber. Until its dismantling, the stone chamber had been preserved at the site by environmental management with the aim of keeping the environment inside and outside the stone chamber in a stable state before excavation. In addition to the generation of mold and other microorganisms as the main problems in the preservation of murals related to the temperature and humidity in the stone chamber, the plaster layer peels off when the stone chamber dries, and the temperature and humidity in the stone chamber must be properly managed. It is necessary to maintain high humidity levels in order to prevent cracking and peeling of the plaster layer. At the same time, to suppress the generation of mold and the like, it is desirable to prevent condensation on the surface of the mural, avoid a high humidity environment, and keep the temperature as low as possible, but considering the above, low humidity levels cannot be achieved. For the latter reason, the relative humidity in the stone chamber was maintained at a high value, close to saturation, as it was before its discovery. Thus, from the construction of the conservation facility to the dismantling phase, using the record of the preservation measures and the measured values of the temperature and humidity in the stone chamber and the estimation results by numerical analysis,

D. Ogura (✉) · S. Hokoi
Graduate School of Engineering, Kyoto University, Kyoto, Japan
e-mail: ogurad@archi.kyoto-u.ac.jp

S. Hokoi
e-mail: hokoi@archi.kyoto-u.ac.jp

S. Hokoi · Y. Li
School of Architecture, Southeast University, Nanjing, China

T. Ishizaki
Institute for Conservation of Cultural Property, Tohoku University of Art and Design, Yamagata, Japan

J. M. P. Q. Delgado (ed.), *Case Studies in Building Constructions*, Building Pathology and Rehabilitation 15, https://doi.org/10.1007/978-3-030-55893-2_2

we examined how the temperature and humidity in the stone chamber changed. The measurements are reported here when possible; in some cases, an estimation through numerical analysis is presented instead.

Keywords Mural painting · Tumulus · Coupled heat and moisture transfer · Conservation facility · Mould

Symbols

As	Absorption rate of solar radiation on the surface
$c\gamma$	Volumetric heat capacity of air (J/m^3K)
$c'\gamma'$	Volumetric moisture capacity of air (kg/m^3Pa)
$(c\rho)_{ap}$	Apparent heat capacity of the material (J/kgK)
g	Acceleration of gravity (= 9.8) (m/s^2)
J_s	Precipitation (kg/m^2s)
M_v	Molecular weight of water (kg/kmol)
N_v	Air change rate between the stone chamber and the outside air (1/s)
n	Vertical downward vector (–)
p	Water vapor pressure (Pa)
p_o	Water vapor pressure of outdoor air (Pa)
$p_{sat}(T_i)$	Saturated water vapor pressure for surface temperature (Pa)
$q_{s(s)}$	Heat flux (W/m^2)
q_{sol}	Absorption of solar radiation (W/m^2)
q_{noc}	Nocturnal radiation (W/m^2)
R_ν	Gas constant of water vapor ($R_\nu = R/M_\nu$)
R	Gas constant (Pa m^3/kmol K)
r	Heat of gas/liquid phase change (J/kg)
S	Surface area of the wall (m^2)
T	Temperature (K)
U	Wind speed (m)
V	Air volume of room (m^3)
α	Heat transfer coefficient (W/m^2K)
α'_m	Moisture transfer coefficient due to water vapor pressure difference (kg/m^2sPa)
α'_T	Moisture transfer coefficient due to temperature difference (kg/m^2sK)
α'_μ	Moisture transfer coefficient due to water chemical potential difference ($kg/m^2s(J/kg)$)
ε_s	Emissivity of the surface (–)
λ	Thermal conductivity (W/mK)
λ'_T	Moisture conductivity related to temperature gradient (kg/msK)
λ'_{Tg}	Vapor conductivity related to temperature gradient (kg/msK)
λ'_μ	Moisture conductivity related to water chemical potential gradient (kg/ms(J/kg))

$\lambda'_{\mu g}$	Vapor conductivity related to water chemical potential gradient (kg/ms(J/kg))
μ	Water chemical potential (J/kg)
ρ_w	Liquid water density (kg/m^3)
ρ	Air density (kg/m^3)
ψ	Volumetric moisture content (m^3/m^3)

Subscript

g	Gas phase
j	The jth wall element
l	Liquid phase
n	Number of wall divisions
o	Outdoor
r	Room
s	Solid surface in contact with air
sat	Saturated

1 Introduction

Takamatsuzuka Tumulus [1] was discovered on March 21, 1972, in Asuka village in Nara Prefecture, as shown in Fig. 1. A stone chamber is buried within the mound of soil. Paintings on the lime wall plaster inside the chamber are believed to have been painted between the end of the seventh century and the beginning of the eighth century. The Takamatsuzuka Tumulus was designated as a special historic site in 1973, and the mural paintings were designated as a national treasure in 1974. It was decided that the mural paintings should be preserved on site to be kept under the same condition as when found. For this purpose, a conservation facility, which consists of three antechambers, was constructed in 1976.

In the spring of 2001, black mold was found at the entrance of the stone chamber after the soil had been consolidated there. In the autumn of the same year, mold was also found inside the stone chamber. In 2005, the Committee for the Conservation of the National Treasure, Takamatsuzuka Tumulus, decided to dismantle the stone chamber in 2007 for restoration, which was safely completed in the same year [2].

The causes of the deterioration of the mural paintings were mainly the growth of mold and microorganisms as well as the exfoliation of the plaster layer by drying. The temperature and humidity in and surrounding the stone chamber affect the factors of deterioration. However, it remains unclear how the variation of climatic conditions and conservation measures influence the temperature and humidity in the stone chamber.

Takamatsuzuka Tumulus

Fig. 1 Position of Nara Prefecture and Takamatsuzuka Tumulus

In this study, measurement data and the simulation of temperature and humidity in the stone chamber in the past are analyzed using records of conservation measures from the construction of the conservation facility to the dismantlement of the stone chamber in order to clarify these deterioration factors.

2 Outline of the Conservation Facility

The conservation facility was built in order to check the mural paintings and stop the exfoliation safely. The facility consists of three antechambers, and the stone chamber connects to the antechamber through a small space, called "Toriaibu" as shown in Fig. 2. Those antechambers are kept under the same condition as when they were found. Temperature control of the antechambers is maintained by heating and cooling panels, which consist of copper tubes in which the water temperature remains at a measured value in the ground of "Toriaibu". Air conditioning maintains the same temperature and almost saturated in humidity in advance when workers enter the stone chamber.

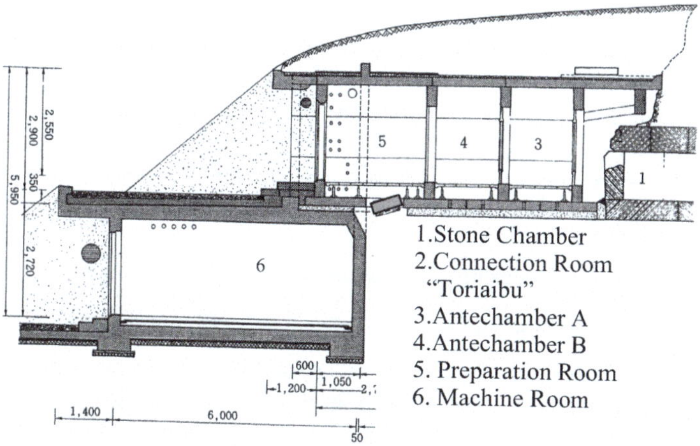

Fig. 2 Section of the conservation facility [1]

3 Analysis of the Measured Temperature in the Stone Chamber

Figure 3 shows the annual average variation of the measured temperature in the stone chamber [3] during the running of the conservation facility. The approximate linear curves obtained through the least squares method are shown in the same figure. The temperature in the stone chamber gradually increased and especially rose from 2001,

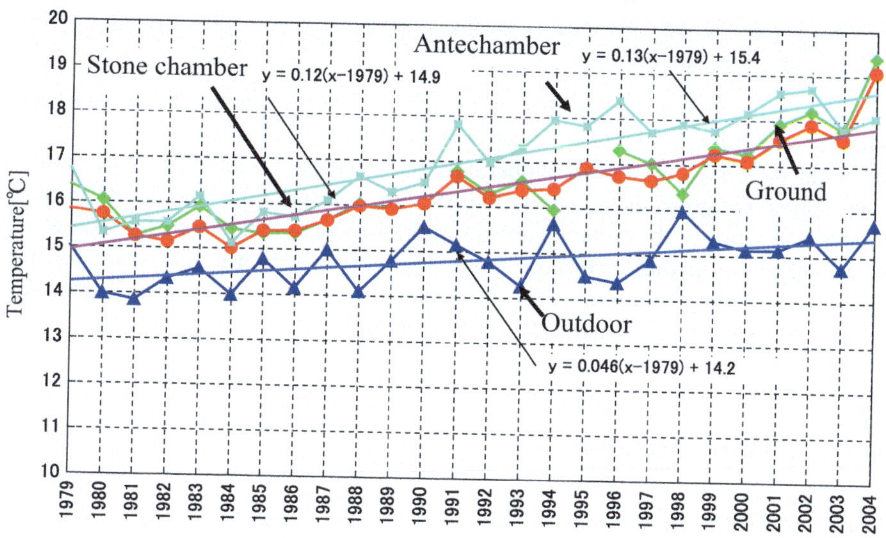

Fig. 3 Annual average variation of the measured temperature

when mold appeared in the stone chamber, and workers entered the stone chamber to remove the mold many times. The slope of the stone chamber temperature is about twice as large as that of the outdoor temperature. The temperature rises using the approximate curve of the stone chamber, antechamber, and outdoor are 2.9, 3.1, and 1.2 degrees, respectively.

The causes of temperature rises in the stone chamber are thought to be as follows.

(1) Change in climatic conditions

- The outdoor temperature rose gradually from 1972 to 2004.

(2) Changes in the mound surface

- Bamboo trees on the mound were cut, and the waterproof sheets were used to cover it to prevent the infiltration of precipitation in 2003.

(3) Temperature control of antechambers

- The temperature, controlled by the heating and cooling panels, was higher than the ground temperature at the highest point in the year for many years.

(4) Heat generation of workers in the stone chamber

- The temperature rose when workers entered the stone chamber.

(5) Temperature of the machine room

- Exhaust heat due to the temperature control of the antechambers was released in the machine room.

(6) Heat conduction through the envelope of the conservation facility

- A part of the envelope is exposed to the southern outside, where solar radiation enters.

We investigated the causes through analysis of the measurement data and simulation results, namely (1), (2), (3), and (4) as well as (6) in all the simulations. We identified three factors causing temperature rises in the stone chamber (i.e., changes in climatic conditions, the effects of temperature control of the heating and cooling panels, and work in the stone chamber) in the following chapter using. The effect of the change in mound surface on the temperature variation in the stone chamber was investigated by simulation. The details are described in the reference, and their effects are taken into account in the simulation [4].

4 Simulation Method [5]

The simulation method is shown in Table 1. The basic equation used in the analysis is the simultaneous heat and moisture transfer equation [21] that deals with the transfer of heat and moisture in the porous material. Many materials, such as mound soil,

Table 1 Calculation condition

Meteorological conditions	Nara meteorological observatory (1975–2005) [10] Outside temperature, outside air relative humidity, horizontal solar radiation[a], rainfall, cloud amount[b]
Physical properties of heat and moisture	Ground: estimated [11] based on the equilibrium water content of rammed earth [12] at Takamatsuzuka Tumulus Sandy soil [13] near the ground surface Tuff: estimated [11] based on literature values [14–16] Stucco: literature values [17] Conservation facility concrete: Literature value [18] Insulation material for conservation facility: extruded foam polystyrene foam reference value [19] is used (non-moisture permeable)
Method of calculation	Explicit finite difference method
Calculation period	January 1, 1979, to December 31, 2005 (Start-up calculation for 4 years from 1975)

[a]The horizontal solar radiation was divided into direct solar radiation (Bouguer's formula) and sky solar radiation (Berlage's formula), and the solar radiation was given considering the angles of the south and north slopes of the mound [20]

[b]The night-time radiation dose was calculated by using Brunt's formula with cloud amount correction [20]

stone room material, and preservation facility frame concrete, are porous materials with voids inside the material, and this equation is applied to the transfer of heat and moisture.

$$(c\rho)_{ap}\frac{\partial T}{\partial t} = \nabla \cdot \left[\left(\lambda + r\lambda'_{Tg}\right)\nabla T + r\lambda'_{\mu g}\nabla\mu \right] \tag{1}$$

$$\rho_w \frac{\partial \psi}{\partial \mu}\frac{\partial \mu}{\partial t} = \nabla \cdot \left[\lambda'_{\mu}(\nabla\mu - g\mathbf{n}) + \lambda'_T \nabla T \right] \tag{2}$$

For room air such as that in a stone room, the following formula is used, with room air being represented by one mass point.

$$c\gamma V\frac{\partial T_r}{\partial t} = \sum_{j=1}^{n} S_j\alpha_j\left(T_j - T_r\right) + c\gamma V N_v(T_o - T_r) + Q_r \tag{3}$$

$$c'\gamma' V\frac{\partial p_r}{\partial t} = \sum_{j=1}^{n} S_j\alpha'_{mj}\left(p_j - p_r\right) + c'\gamma' V N_v(p_o - p_r) + J_r \tag{4}$$

The relationship between water chemical potential and water vapor pressure is obtained from the following equation.

$$\mu = R_v T\,\ln\left(p/p_{sat}\right) \tag{5}$$

Boundary conditions on the ground surface in contact with the outside air include the transpiration from the bamboo forest. The bamboo forest is a simple system in which the plant layer and the ground surface layer are combined and treated as a single layer. This model considers solar radiation absorption and the transpiration of plants, and the degree of their influence on the stone room environment can be sufficiently examined. The equations for the heat and moisture boundary conditions and the transpiration amount on the ground surface in contact with the outside air are shown below.

The thermal boundary condition of the ground surface in contact with outside air is as follows.

$$\left(\alpha + b \cdot r\alpha'_T\right)(T_o - T_s) + b \cdot \alpha'_\mu(\mu_o - \mu_s) + A_s q_{sol} - \varepsilon_s q_{noc} - rE \\ = -\left(\lambda + r\lambda'_{Tg}\right)\left.\frac{\partial T}{\partial n}\right|_s - r\lambda'_{\mu g}\left.\frac{\partial \mu}{\partial n}\right|_s \tag{6}$$

The boundary condition of moisture on the ground surface in contact with outside air is as follows.

$$b \cdot \left[\alpha'_\mu(\mu_o - \mu_s) + \alpha'_T(T_o - T_s)\right] + J_s = -\lambda'_\mu\left(\left.\frac{\partial \mu}{\partial n}\right|_s - g\mathbf{n}\right) - \lambda'_T\left.\frac{\partial T}{\partial n}\right|_s \tag{7}$$

The amount of evaporation from the vegetation [6], which is caused by the difference between the saturated water vapor pressure at the leaf surface and the water vapor pressure of the surrounding air is as follows.

$$E = a\rho\beta C_H U(p_{sat}(T_s) - p_o)$$

Here, a is the ratio of the area of plants (bamboo forest) when the area of the ground surface is 1 as a parameter of transpiration, b denotes the ratio of the area of no plants when the area of the ground surface is 1, β represents the evaporation efficiency, and C_H is the bulk transfer coefficient. The evaporation area from the ground is 30% compared to when the mound is bare land. It should be noted that the water vaporized is assumed to be uniformly removed from the entire mound for convenience. As shown in Fig. 4, the analytical model is treated as a two-dimensional system in which the north-south cross section of the stone chamber is taken out. In the conservation facility, the temperature of the anterior room was controlled to be equal to the temperature in the stone room according to the design policy [1], and there was no exhaust heat from the machine room or entry (by workers) into the stone room. Table 1 shows the meteorological conditions and thermophysical properties used in the calculations.

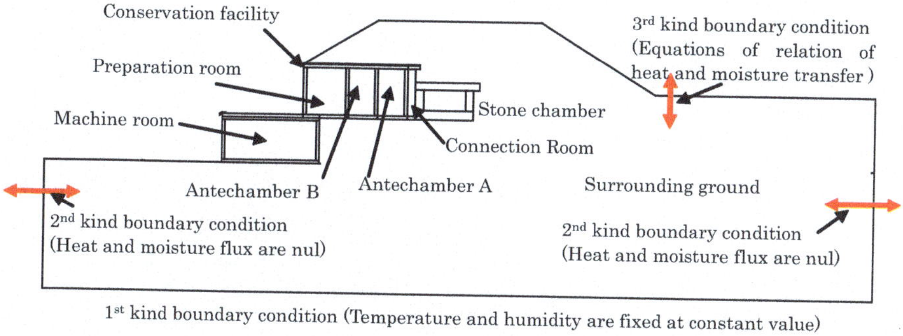

Fig. 4 Schematic diagram for analysis

5 Analysis of the Causes of the Temperature Rise in the Stone Chamber

We mainly investigated causes of temperature rise in the stone chamber. We picked up 3 factors which affect temperature rise of the stone chamber, hence change of climatic condition, effect of temperature control of heating and cooling panel and work in the stone chamber as follows. Change of the mound surface is investigated in the reference [4].

5.1 Investigation of Effects of Weather Conditions [5]

5.1.1 Calculation Conditions

Table 2 shows the setting conditions for conservation facility and so on.

Table 2 Setting conditions such as conservation facility and so on

Temperature control in the front room and the preparation room	Machine room	Stone chamber	Mound surface
1. Controlled to be equal to the stone chamber temperature 2. No ventilation with the mating section	1. No heat or moisture generation 2. No ventilation with outside air	1. No heat or moisture generation 2. No ventilation between the stone chamber and the joint	1. Vegetation is evenly distributed 2. Coating conditions do not change with time

Fig. 5 Comparison of measured and calculated temperature fluctuations in the stone chamber (1979–2005)

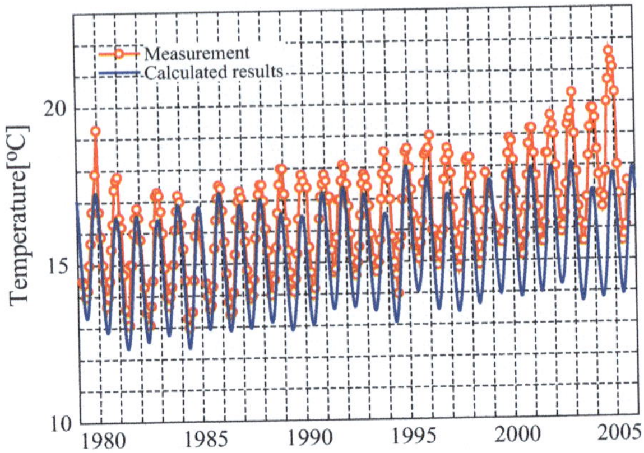

5.1.2 Results and Discussions

Figure 5 shows a comparison of the analysis results and the measurement results of temperature fluctuations in the stone chamber. The figure indicates that the analysis value of the temperature in the stone room underwent a moderate increase with respect to the temperature increase of the measured value for 26 years in all periods. Therefore, it can be concluded that changes in meteorological conditions constitute one of the main factors of the temperature rise during this period, but they cannot completely explain the temperature rise.

5.2 *Effect of Temperature Control of Heating and Cooling Panel [7]*

5.2.1 Calculation Conditions

Table 3 shows the actual changes in the covering conditions of the mounds [8].

If the forest was covered with bamboo forest, a vegetation model was employed, which is divided into two layers—the bamboo forest and the ground surface—where

Table 3 Changes in the coverage of mounds

Period	Cover condition of mounds
1976–September 2003	Bamboo forest covers the surface of the mound
September 2003–2004	Bamboo on the top of the mound is felled and a waterproof sheet is installed
September 2004	Bamboo in the lower part of the mound is felled, the waterproof sheet in the upper part of the mound is removed, and a temporary cover for excavation is constructed

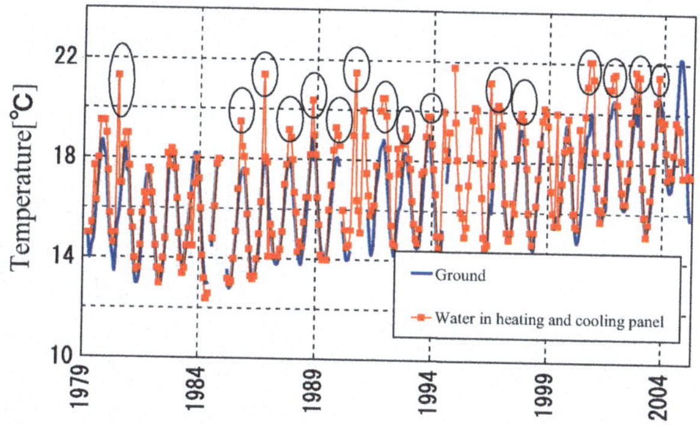

Fig. 6 Measured temperature of the inlet water of the heating and cooling panel

the heat and moisture resistance of the vegetation and solar shading could be considered. For the waterproof sheet, the solar reflectance of the sheet was measured [4]. The waterproof sheet was installed at the top of the mound, referring to the drawings and photographs at the particular time, and a temporary cover was installed by removing the waterproof sheet in the same area as the sheet. Using the recorded value [3] for the temperature of the water supplied to heating and cooling panel, the temperature and humidity of the anteroom, ground, and stone chamber were calculated, assuming that the heating and cooling panel would control the preparation room and the antechamber. The water temperature of the heating and cooling panel used as input in the analysis is shown in Fig. 6, along with the antechamber and underground temperature. The water temperature was supposed to be controlled to be equal to the underground temperature, but in many years, it was higher than the underground temperature (circled).

The temperatures of the heating and cooling panel were higher than those in the ground in November and December in many years from 1979 to 2004. A simulation was performed using the measured temperature of the panel.

5.2.2 Results and Discussions

Figure 7 shows a comparison between measured and calculated temperatures in the stone chamber. Changes in the outdoor climate and the mound are considered in Case 1. In Case 2, we take into account changes in the outdoor climate and the mound as well as controls for the antechamber making use of the measured temperature of the heating and cooling panel. The calculated room temperature in the stone chamber in Case 2 corresponds closely with the measured room temperature in the stone chamber except for the period from 2002 to 2004. Therefore, a calculation considering the climatic change and the measured temperature variation of the heating and cooling panel explains, more or less, the temperature rise in the stone chamber for 30 years. A calculation considering the conditions described above cannot explain the temperature variations in the stone chamber from 2002 to 2004.

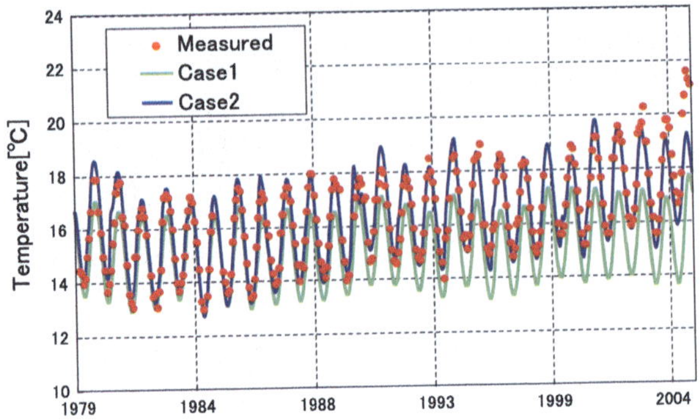

Fig. 7 Comparison between measured and calculated temperature

5.3 *Effect of Heat and Moisture Generation Due to Work in the Stone Chamber [7, 9]*

5.3.1 Calculation Conditions

Figure 8 shows the measured temperature variations in the stone chamber when several hours of work in the stone chamber continued for five days before the stone chamber was dismantled in 2006. Since the temperatures in the stone room are not measured by the sensor when entering the room, the data when entering the stone chamber were omitted. The figure illustrates that a temperature rise accumulated day by day during the research period. We calculated the following conditions to investigate the effect of working in the stone chamber on temperature rises.

1. The copper tube surface temperature was controlled via the inlet water temperature, which is set at the ground temperature of "Toriaibu".

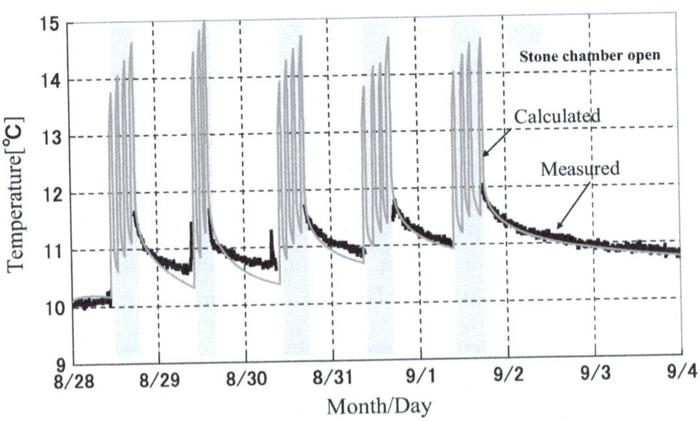

Fig. 8 Temperature in the stone chamber during 5 days of work (August 28–September 3, 2006)

2. The working condition was provided by Prof. Masuda, who restored the mural paintings on site at that time, and we also used his working diary and other books.

We calculated two cases: work in the stone chamber and no work in the chamber.

5.3.2 Results and Discussions

Using the measured values of temperature and humidity in the anterior chamber at the time of continuous entry and exit from the stone chamber immediately before the dismantling of the stone chamber, the generation of heat and moisture in the internal work was estimated and analyzed. In Fig. 8, the analysis value of the temperature in the stone room can sufficiently reproduce the measured value of the temperature change after leaving the room. Figure 9 shows the comparison of the case with and without work being performed. During work periods, the temperature varied more than non-work periods; work to restore the painting continued from 1979 to 1981, and a period consisting of frequent checks occurred after 2001. The results can be summarized as follows. After continuous work in the stone chamber, the temperature in the stone chamber became slightly higher than before; this is one of the main causes of the temperature rise in the stone chamber.

In addition, as shown in Fig. 10, the relationship between the conditions of entry into the stone chamber, the temperature in the stone chamber, and the occurrence of mold in the stone chamber is that the temperature in the stone chamber is high, and the frequency of the occurrence of mold is high when many people enter the rooms. These elements are likely to be correlated.

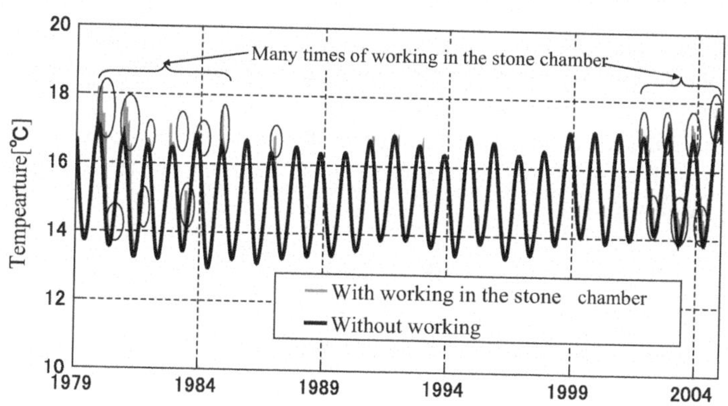

Fig. 9 Comparison between the results of work and non-work in the stone chamber

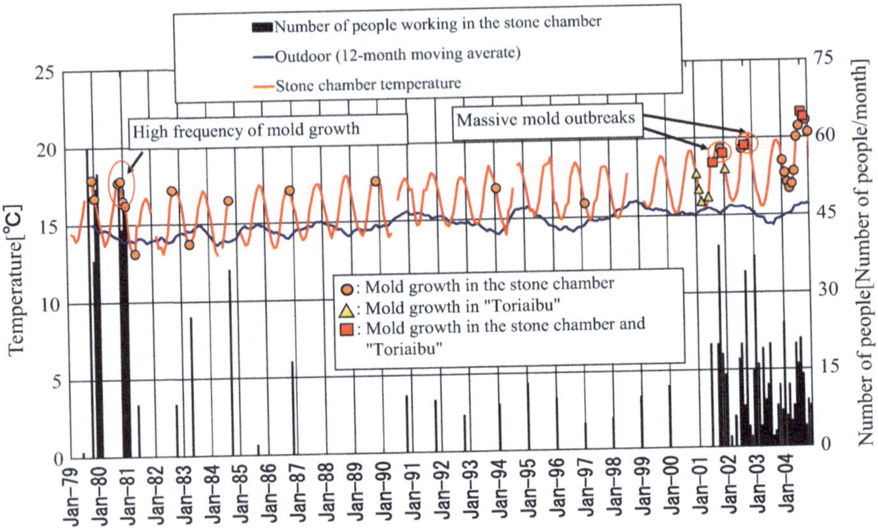

Fig. 10 Mould occurrence, number of people in the room, and stone chamber temperatures

5.4 *Investigation of Change in the Mound Surface [4]*

To properly evaluate changes in the covering conditions through numerical analysis, we again introduced a model considering the solar radiation transmittance of plants as well as heat and moisture resistance and used the measured value as the solar reflectance (albedo) of the waterproof sheet. The results obtained are as follows:

1. Two of the causes of the temperature rise in the stone chamber around the end of 2004 are possibly the felling of the bamboo grove and the installation of a waterproof sheet in September 2003.
2. The installation of a temporary shed in September 2004 was probably one of the factors that lowered the temperature in the stone room rather than continuing to use the waterproof sheet.

5.5 *Influence of Machine Room Temperature on Stone Chamber Temperature and Humidity [7]*

An example of the temperature measurement value of the machine room is shown in Fig. 11. The measurements were taken only for one year during the period when the ground was cooled and the stone chamber was cooled to 10 °C prior to the dismantlement of the stone chamber, but the temperature was 25–30 °C throughout the year, which is high, compared to the ground temperature. This is due to the heat exhausted to the machine room by cooling.

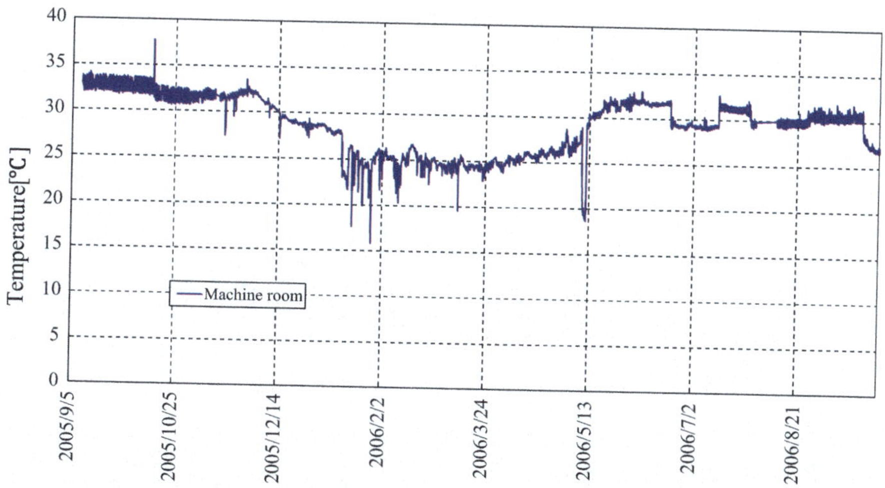

Fig. 11 Machine room temperature before dismantling the stone chamber

The results were obtained under the condition that the cooling rate was significantly increased from before the dismantlement because the ground was cooled to 10 °C at the same time as it was cooled from the ante chamber. It is unclear if past conditions have resulted in such a hot environment, but the temperature of the ground around the stone chamber may have increased due to exhaust heat from the machine room.

5.6 Influence of Heat Transfer in the Body of Conservation Facility on Temperature and Humidity in the Stone Chamber [7]

Figure 12 shows the measured values of the temperature of the preparation room, the front room, and the stone chamber from 1979 to 2005. The temperature of the preparation room is higher than the front room temperature throughout the year. This may be due to the effects of solar radiation on the south side of the preparation room and the high machine room temperature under the preparation room. Since there is no vegetation on the south side of the preparation room and the effect of rainfall is small, it is highly possible that the temperature remains high throughout the year because of solar radiation. In the analysis results thus far, heat transfer through the abovementioned structure of the conservation facility has been considered.

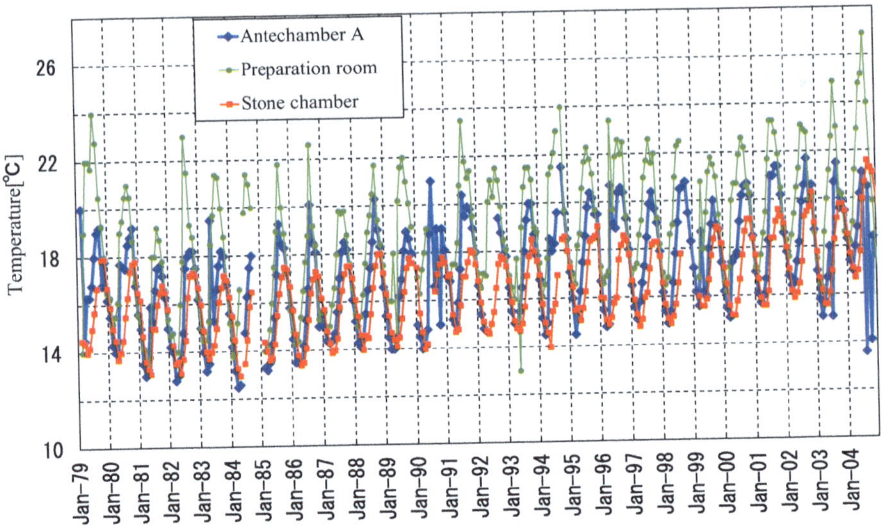

Fig. 12 Preparation room, antechamber, stone chamber temperature measurements

6 Conclusions

In this study, we clarified some factors related to how variations in the climatic conditions as well as the conservation measures implemented, influence the temperature and humidity in the stone chamber of Takamatsuzuka Tumulus in the period when the conservation facility was running from 1976 to 2004 before the dismantlement of the stone chamber, making use of a simulation and measured results. The obtained main results are as follows.

1. Changes in meteorological conditions constitute one of the main factors of the temperature rise during this period, but they cannot completely explain the temperature rise in the stone chamber.
2. A calculation considering the climatic change and the measured temperature variation of the heating and cooling panel explains, more or less, the temperature rise in the stone chamber for 30 years. A calculation considering the conditions described above cannot explain the temperature variations in the stone chamber from 2002 to 2004.
3. After continuous work in the stone chamber, the temperature in the stone chamber became slightly higher than before; this is one of the main causes of the temperature rise in the stone chamber.
4. The relationship between the conditions of entry into the stone chamber, the temperature in the stone chamber, and the occurrence of mold in the stone chamber is that the temperature in the stone chamber is high, and the frequency of the occurrence of mold is high when many people enter the rooms. These elements are likely to be correlated.

5. Two of the causes of the temperature rise in the stone chamber around the end of 2004 are possibly the felling of the bamboo grove and the installation of a waterproof sheet in September 2003.
6. The installation of a temporary shed in September 2004 was probably one of the factors that lowered the temperature in the stone room rather than continuing to use the waterproof sheet.
7. The temperature of the ground around the stone chamber may have increased due to exhaust heat from the machine room.
8. Since there is no vegetation on the south side of the preparation room and the effect of rainfall is small, it is highly possible that the temperature remains high throughout the year because of solar radiation.

In the simulation, we used measurement data, records, and physical properties as thoroughly as possible, but we assumed many factors. Thus, we think that these results reflect the causes of the temperature increase the conditions at the time.

Acknowledgements We deeply appreciate Prof. Katsuhiko Matsuda and Mr. Sadatoshi Miura, who gave us advice and comments about countermeasures and work at the time. This research was partially supported by the Ministry of Education, Science, Sports, and Culture, Grant-in-Aid for Scientific Research (A), 18206062, 2006, 2007 as well as Grant-in-Aid for Scientific Research (C), 20560549, 2008–2010.

References

1. Agency for Cultural Affairs: National Treasure Takamatsuzuka Tumulus Paintings–Conservation and Repair (1987) (in Japanese)
2. Agency for Cultural Affairs: Overview of Stone Chamber Dismantlement Project of Takamatsuzuka Tumulus (2007) (in Japanese)
3. T. Ishizaki, Temperature and Humidity in the Stone Chamber of Takamatsuzuka Tumulus in the Past, Takamatsuzuka Tumulus Mural Painting Deterioration Cause Investigation Committee, Agency for Cultural Affairs, 4th meeting (2008) (in Japanese)
4. D. Ogura, S. Hokoi, Y. Li, T. Ishizaki, Analysis of temperature and humidity variation in the stone chamber of takamatsuzuka tumulus from its discovery until its dismantlement: part 2—effect of different mound covering conditions during the operation of the conservation facility. Conserv. Sci. (49), 73–85 (2009) (in Japanese)
5. D. Ogura, S. Hokoi, Y. Li, T. Ishizaki, S. Miura, Analysis of temperature and humidity variation in the stone chamber of takamatsuzuka tumulus from its discovery until its dismantlement—the effect of outside air temperature change during the operation of the conservation facility, and the effect of the temporary shelter set up immediately following excavation. Conserv. Sci. **48**, 1–11 (2009) (in Japanese)
6. J. Kondo, Editing and writing: meteorology of water environment—water balance and heat balance on the ground surface. (Asakura Shoten, 1996), pp. 208–239 (in Japanese)
7. D. Ogura, S. Hokoi, Y. Li, T. Ishizaki, Analysis of temperature and humidity variation in the stone chamber of takamatsuzuka tumulus from its discovery until its dismantlement: part 3—effect of temperature of the heating and cooling panel and work in the stone chamber during the operation of the conservation facility. Conserv. Sci. (49), 87–96 (2009) (in Japanese)

8. Agency for Cultural Affairs, National Treasure Takamatsuzuka Tumulus Mural Paintings Preservation History (1972–September 2006). https://www.bunka.go.jp/seisaku/bunkazai/takamatsu_kitora/hozon_kako.html (in Japanese)
9. D. Ogura, S. Hokoi, T. Ishizaki, Y. Li, H. Akasaka, Conservation environment of mural paintings in tumulus Part3: Effect of temperature of the heating and cooling panel and work in the stone chamber during the operation of the conservation facility, in *Architectural Institute of Japan Kinki Branch Research Report*, vol. 50, pp. 201–204, 2010 (in Japanese)
10. Japan Meteorological Business Support Center: Meteorological database Ground observation, CD-ROM, (2007)
11. Y. Li, D. Ogura, S. Hokoi, T. Ishizaki, Effects of emergency preservation measures following excavation of mural paintings in Takamatsuzuka Tumulus. J. Building Phys. **36**(2), 117–139 (2012)
12. M. Khalil, T. Ishizaki, Moisture characteristic curves of the soil of takamatsuzuka tumulus. Conserv. Sci. **46**, 13–20 (2006)
13. W. Jury, *Simultaneous Transport of Heat and Mass Through a Medium Sand*, Ph.D. Thesis (University of Michigan, 1973)
14. W. Lin, M. Takahashi, D. Sato, E.-C. Yeh, Y. Hashimoto, W. Tanikawa, Measurement of rock pore size distribution by mercury injection method. J. Japan Soc. Eng. Geol. **57**(5), 201–212 (2016)
15. K. Kodai, Graphic representation of rock permeability. Bull. Geol. Surv. Jpn. **35**(9), 419–434 (1984) (in Japanese)
16. K. Kitano, K. Shin, N. Kinoshita, T. Okuno, Mechanical, thermal properties and permeability of rocks under high temperature. J. Jpn. Soc. Eng. Geol. **29**(3), 36–47 (1988) (in Japanese)
17. Architectural Institute of Japan: Building Design Data Collection I Environment, Maruzen, p. 176 (1978)
18. D. Ogura, S. Mino, T. Matsushita, M. Mizuhata, M. Matsumoto, Analysis of heat and moisture behavior in the underground space and its surrounding ground. J. Architectural Plann. Environ. Eng. Architectural Inst. Jpn. (539), 15–21 (2001) (in Japanese)
19. Japan Society for Thermophysical Properties, in *Handbook of Thermophysical Properties* (2000), p. 218 (in Japanese)
20. S. Tanaka, H. Takeda, et al., *Latest Architectural Environmental Engineering* 2nd edn. (Inoue Shoin 1999), pp. 91–99, (in Japanese)
21. M. Matsumoto, New series of architecture, in *Moisture*, vol. 10 (Shokokusha Co. Ltd., 1984) (in Japanese)

Estimation of Seismicity Parameters and a Backpropagation Neural Network for Prediction of Earthquake Magnitude in Northeast Region of India

Amit Zarola and Arjun Sil

Abstract This study presents the spatial variation of seismicity parameters and artificial neural network (ANN) model for earthquake magnitude prediction in the northeast region of India considering updated earthquake catalogue of magnitude Mw $\geq$ 4.0 that occurred from year 1731 to 2015 in the study area. The study area has been divided into 17 seismic source zones keeping in view the spatial variation in earthquake occurrences, distribution pattern of events and orientation of seismic sources. Seismicity parameters were estimated for each source zone as well as for the whole study area. The estimated '*b*' value in this study varies from 0.50 to 0.84 in different seismic source zones and for the whole study area it is found as 0.77. The proposed ANN technique to predict the possible magnitude of earthquakes in the identified seismic source zones is based on feedforward backpropagation neural network model with single hidden layer. Total five input parameters namely, longitude, latitude, elapsed time between two events, cumulative magnitude probability and seismic energy and one output parameter namely, magnitude of earthquakes were used in ANN. The '*b*' value estimated in this study was used as an input unit in ANN in calculating the cumulative magnitude probability for different zones. The performance of ANN was evaluated by estimating the mean absolute error (MAE), sum of squared error (SSE) and Mean squared error (MSE). And the results obtained in this study show that the ANN model yields good prediction accuracy for earthquake magnitude in NE India.

Keywords Seismicity parameters · ANN · Seismic source zones · Cumulative magnitude probability · Seismic energy

A. Zarola · A. Sil (✉)
Dept of Civil Engineering, NIT Silchar, Silchar, Assam 788010, India
e-mail: silarjun@gmail.com

J. M. P. Q. Delgado (ed.), *Case Studies in Building Constructions*, Building Pathology and Rehabilitation 15, https://doi.org/10.1007/978-3-030-55893-2_3

1 Introduction

The general property of the size distribution of earthquakes that smaller earthquakes occur more frequently than the bigger earthquakes is well known. Gutenberg and Richter [1] proposed the linear relationship between the magnitude and frequency of the earthquake. It is named as Frequency-Magnitude Relation and it is given as:

$$\log_{10}\mathrm{N(m)} = \mathrm{a} - \mathrm{bm} \quad (1)$$

where m is the magnitude of earthquake and $N(m)$ is the cumulative number of events with magnitude higher than m or can be defined as the mean annual rate of occurrence of a particular magnitude m and above. 'a' and 'b' are defined as seismicity parameters. a-value describes the seismic activity (Log number of seismic events with magnitude = 0), higher value of 'a' indicates the higher level of seismicity and b-value is a tectonic parameter which shows the slope of the straight line, higher value of 'b' indicates a very high proportion of smaller earthquakes to larger earthquakes, small changes of 'b' value provides much difference in energy released.

However, the 'a' and 'b' value are the most widely reported statistics in the seismic hazard assessment and earthquake prediction. Several studies were performed in estimation of frequency-magnitude relationship and spatial & temporal variations of its seismicity parameter (b-value) in the different regions of the world [2, 3]. Indeed, Frohlich and Davis [4] found globally averaged b-value as approximately 1.0. Laboratory studies show that there are some natural factors which can cause significant changes in b-value from an average value 1.0. For example, Mogi [5] shows an increase in the heterogeneity of material causes increase in b-value while Warren and Latham [6] show that an increase in the thermal gradient may cause an increase in the b-value.

There are papers in the literature that made attempts to predict the magnitude and intensity of earthquake using ANN technique [7–10]. However, Wong et al. [7] have presented Modified Mercalli Intensity (MMI) forecast model for use in regional seismic hazard assessment in California using ANN method. The authors have used three-layer backpropagation neural networks with hyperbolic tangent sigmoid transfer function and normalized cumulative-delta learning rule. On the other hand, Alarifi et al. [8] proposed the application of ANN to predict the magnitude of future earthquakes in northern Red Sea area using different feedforward neural network configurations with multi-hidden layers along with two transfer functions namely log-sigmoid and hyperbolic tangent sigmoid and the performance of the neural networks was evaluated by estimating the MAE and MSE.

Moustra et al. [9] performed two different case studies to evaluate the performance of ANN in predicting the earthquake occurrence in the region of Greece using feedforward backpropagation multilayer perceptron type neural network. Similarly, Narayanakumar and Raja [10] also presented a three layer feedforward backpropagation neural network model to predict magnitude of earthquakes in the region of Himalayan belt. Location, energy released and seismicity parameters have been taken as input elements and magnitude as an output element to prepare the architecture of neural network. Many other researchers performed the same study in the different regions of the world and suggested that ANN to be the most powerful tool in forecasting the magnitude and occurrence of earthquakes.

In the present study, the northeast region of India is considered as study area and divided into 17 identified seismic source zones on the basis of distribution pattern of seismic events of latest catalogue and orientation of seismic sources or faults. The earthquake events were collected since 1731–2015 from available international and national various seismological agencies such as United State Geological Survey (USGS), National Geographical Research Institute (NGRI), Indian Meteorological Department (IMD) and Bhabha Atomic Research Centre (BARC). There are mainly two goals of the present study. The first goal of this study is to estimate the seismicity parameters ('*a*' and '*b*' value) for these 17 seismic source zones in the northeast region of India and the second goal of this study is to investigate the performance and accuracy of ANN in forecasting the magnitude of earthquakes in the northeast region of India. Hence, the neural networks were trained for input and output data of earthquake events occurred from year 1731 to 1990 and tested for input data of earthquake events occurred from year 1991 to 2015. The forecasted magnitudes in the years 1991 to 2015 were compared with the actual magnitudes by estimating MAE, SSE, and MSE. The proposed Artificial Neural Network (ANN) technique is based on feedforward backpropagation neural network model with single hidden layer.

2 Status of the Study Area and Seismotectonic Setting

The India plate in the north is subducting under Eurasian plate and in the east, it is subducting and colliding under Burmese plate [11–14]. Therefore, the Burma plate consisting of Indo Burmese wedge, Myanmar Central Basin along with Andaman Sumatra arc is rotated clockwise [13]. Actually Burma plate is squeezed between the India plate to the west and Sunda plate to the east. Due to these complex tectonic settings, various major earthquakes were reported in the past.

In the present research work the study area is bounded between latitudes 19.345° to 29.431° and longitudes 87.590° to 98.461°. This region is situated in the zone-V with a zone factor 0.36 g on the latest version of seismic zoning map of India, given in the earthquake resistant design code of India [15] which expects the highest level of seismicity in the country. Many geologically active faults were identified in this region whose activities make the region seismically vulnerable and cause huge destruction of buildings and other structures. Within this study area, total 29 seismic sources including linear sources, thrusts, lineaments and area sources have been identified from different literatures, SEISAT 2000 and remote sensing images. Hence, the study area was divided into 17 seismic source zones (Fig. 1) on the basis of the distribution pattern of seismic events and orientation of seismic sources.

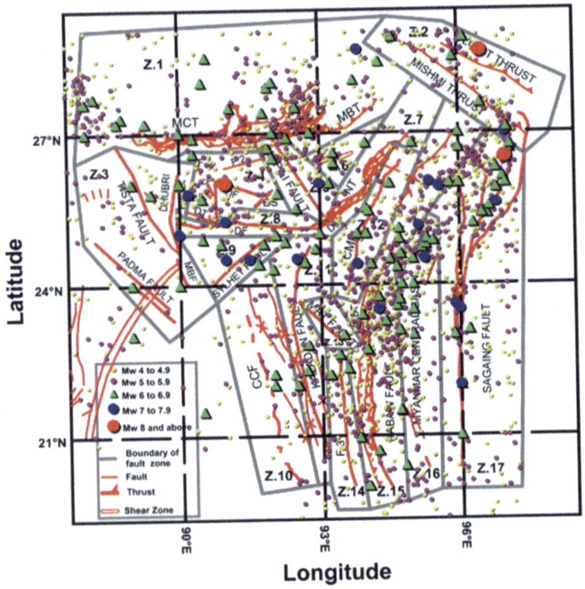

Fig. 1 Delineated 17 identified seismic source zones in northeast India with seismic sources: Main central thrust (MCT), Main boundary thrust (MBT), Lohiti thrust, Mishmi thrust, Tista fault, Padma fault, Dhubri fault, Madhupur blind fault (MBF), Oldham fault (OF), Barapani shear zone (BS), Dapsi thrust (DT), Samin fault (SF), Dudhnoi fault (DhF), Chedrang fault (CF), F.1, F.2, Dauki fault (DF), Kopili fault, Naga thrust (NT), Disang thrust (DsT), Churachandpur Mao fault (CMF), Chittagong coastal fault (CCF), Kaladan fault, Mat fault, Eastern boundary thrust (EBT), Kabaw fault, Myanmar central basin (MCB), Sagaing fault. The earthquake distribution (Mw $\geq$ 4.0) collected from year 1731–2015 is also shown in the figure

3 Earthquake Catalogue

In the present study, the earthquake events of magnitude Mw $\geq$ 4.0 (since 1731–2015) have been considered to estimate the seismicity parameters in the study area. A well assessed earthquake catalogue has been taken from Sil et al. [16], from year 1731 to 2011 (280 years) in the study area within East longitudes 87.590° to 98.461° and North latitudes 19.345° to 29.431°. A complete earthquake catalogue from year 1731 to 2015 has been prepared after adding two earthquake catalogues, first from year 1731 to 2011 [16] and second from year 2012 to 2015 (prepared in this study). The earthquake events from year 2012 to 2015 were collected in the study area from available international seismological agency USGS. To prepare the earthquake catalogue from year 2012 to 2015 same approaches were followed as followed by Sil et al. [16]. First, all the repeated events were removed from the data base and then different types of magnitude scales were converted into a common magnitude scale i.e. moment magnitude (Mw) using correlations proposed by Sitharam and Sil [17].

$$\mathrm{Mw} = 0.862\mathrm{Mb} + 1.034 \quad \left(\mathrm{R}^2 = 0.782\right) \tag{2}$$

$$\mathrm{Mw} = 0.673\mathrm{Ml} + 1.730 \quad \left(R^2 = 0.852\right) \tag{3}$$

$$\mathrm{Mw} = 0.625\mathrm{Ms} + 2.350 \quad \left(\mathrm{R}^2 = 0.894\right) \tag{4}$$

where Mw is the moment magnitude, Mb is the body wave magnitude, Ml is the local magnitude and Ms is the surface wave magnitude. After that the earthquake catalogue was declustered using methodology given by Gardner and Knopoff [18] and modified by Uhrhammer [19]. This methodology used to remove foreshocks and aftershocks assumes that the temporal and spatial distribution of foreshocks and aftershocks are dependent on the size of the main event. The size of space and time window to remove foreshocks and aftershocks are given as $e^{-1.024+0.804*Mw}$ (km) and $e^{-2.87+1.235*Mw}$ (days) respectively.

After adding these two catalogues total 3526 main shock events were obtained among which 2508 events were of magnitude Mw $\geq$ 4.0 within the study area. Hence, total 2508 earthquake events (Mw $\geq$ 4.0) were compiled in the data base. A histogram of earthquake data (year 1731 to 2015) in the study area was prepared (Fig. 2).

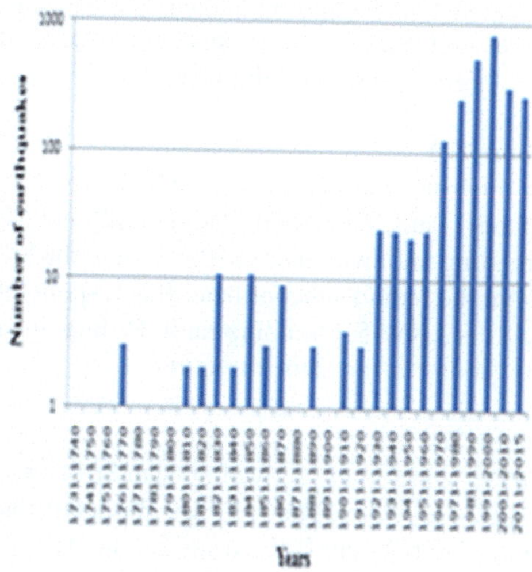

Fig. 2 The figure shows a histogram of earthquakes (year 1731-2015) in the study area

4 Identified Seismic Source Zones in the Study Area

However, considering the seismicity pattern, epicentre distribution of past events and orientation of faults, following zones were identified as the most active source zones in the region for further study as:

Zone Z.1

In this zone, the Eastern Himalayan range is characterized by number of north heading thrusts. Main boundary thrust (MBT), main central thrust (MCT) and main frontal thrust (MFT) are very important amongst them. The entire NE Himalayan belt shows transverse tectonics [20]. This region is seismically very active and responsible for 1947 Arunachal Pradesh earthquake (Ms = 7.7) [14] and recent September 17, 2011 Sikkim earthquake (Mw = 6.9) [14].

Zone Z.2

This zone is known as zone of Assam syntaxis. Mishmi thrust, Lohiti thrust and end part of the Disang thrust is present in the syntaxis zone which is the meeting place of the Himalayan and the Indo-Burmese arc. Molnar [21], Seeber and Armbruster [22] suggested that ongoing subduction of India plate under the Eurasian plate causes great and major earthquakes in this zone. This Assam syntaxis is responsible for 1950, great Assam earthquake.

Zone Z.3

In this zone, mainly four active faults were identified and delineated in Fig. 1, namely Dhubri fault, Tista fault, Padma fault and Madhupur blind fault. Dhubri fault is N-S trending fault which lies to the west of the Shillong plateau and separating the plateau from the Indian subcontinent. It is responsible for 1930 Dhubri earthquake. The tista fault is a NW-SE trending fault. Padma lineament and madhupur blind fault are also traversing in the same direction.

Zone Z.4

In this zone, many active faults and fractures are present such as Oldham, Chedrang, Samin, Dudhnoi, Dapsi thrust, Barapani shear zone, Bordwar fracture and Brahmaputra fault. As per Bilham and England [23], the Oldham fault is a steep, ESE striking fault dipping SSW at 57° present in the northern part of the Shillong plateau. Length of Oldham fault is 110 km which demarcates the northern boundary of the Shillong plateau and it is responsible for 1897 earthquake (Mw = 8.1) [14] as suggested by Bilham and England [23], Nayak et al. [24] and Saha et al. [25]. But Rajendran et al. [26] suggested that the great earthquake of 1897 had occurred on the Brahmaputra fault and it is the actual northern boundary of the Shillong plateau. The 20 km long NW-SE trending Chedrang fault, 4 km long E-W trending Samin fault and 11 km long Bordwar fracture were developed after the 1897 Shillong earthquake near the Shillong plateau [26].

Baruah et al. [27] studied the tectonics of the Chedrang valley and its vicinity area in the western part of the Shillong plateau. They suggested the tectonics of this region is mainly influenced by NW-SE oriented Chedrang fault, N-S oriented Dudhnoi fault, NW-SE oriented Samin fault and E-W oriented Dapsi thrust. The Dapsi thrust is the extension of the Dauki fault along NW-SE direction. It is 90–100 km long reverse fault with strike slip component.

Zone Z.5

In this zone, Kopili fault is present. The Kopili fault zone is also a main intra plate fault zone in the region as Shillong plateau which transgresses into the Himalayan up to the MCT. Kayal et al. [28] and Battacharya et al. [29] suggested that Kopili fault zone is approximately 300 km long and 50 km wide zone which is like a divider between Shillong plateau and Mikir massif and it is a north dipping strike slip fault. Nandy [30] suggested that 1869 Cachar earthquake (Mw = 7.5) [31, 32] occurred on south eastern end of Kopili fault zone. Kopili fault is also responsible for 1943 Assam earthquake (Ms = 7.2) [14].

Zone Z.6

The area between Assam syntaxis and Mikir hills is seismically less active and named as "Assam Gap" [33] and "Aseismic Corridor" [34]. They suggested that a large earthquake may be expected in this area.

Zone Z.7

In this zone, Naga and Disang thrusts are present. The eastward subducting India plate under Eurasian plate caused formation of Indo-Burma ranges which comprises the Naga thrust and the Arakan-Yoma fold belt. The extension of the Naga thrust to the southwest is named as Disang thrust which extends as haflong fracture zone and then it joins the east-west trending Dauki fault [35].

Zone Z.8

In this zone, Dauki fault is present. Approximately 320 km long EW trending Dauki fault demarcates the boundary between Shillong plateau and Bengal basin. Nandy [30] suggested that Dauki fault is a near vertical or a south dipping strike slip/normal fault. 1923 Meghalaya earthquake (Ms = 7.1) [14] occurred on Dauki fault. Among the historic earthquakes the 1548 earthquake, the 1664 earthquake and 1897 earthquake occurred due to the activities of the Dauki fault [36].

Zone Z.9

In this zone, Sylhet fault is present. NE-SW trending Sylhet fault is the active fault in the Bengal basin. The 1918 Srimangal earthquake (Mw = 7.6) occurred beneath the Bengal basin due to the rupture along the Sylhet fault.

Zones Z.10, Z.11 and Z.13 to Z.16

In these zones, the Chittagong coastal fault (CCF), Kaladan, Mat, Myanmar central basin (MCB), Eastern boundary thrust (EBT), Kabaw fault and other fractures are present and delineated in Fig. 1. These are the major faults which affect the Indo Burmese wedge by strike slip faulting. Maurin and Rangin [37] proposed that CCF is a new major fault that onset after the Kaladan fault due to the westward progression of the Indo Burmese wedge. The CCF is seismically less active but had experienced two major earthquakes in 1851 and 1865. While Sikder and Alam [38] described that the Kaladan fault bounds the outer Indo Burmese wedge to the east. The Lelon fault is in between the metamorphic core of the range and the accretionary wedge and the Kabaw fault is in between the Indo Burmese wedge and Myanmar central basin.

Zone Z.12

In this zone, Churachandpur-Mao fault (CMF) is present. CMF is an N-S trending strike slip fault which is responsible for low magnitude earthquakes. Kumar et al. [39] found 0.5 mm/year slip rate in southern part and 3.9 mm/year slip rate in northern part of the CMF by geodetic observations during year 2004–2005 and suggested that the southern part of CMF is less active while the northern part is deformed at higher rates. The January 4, 2016 Manipur earthquake ($Mw = 6.7$) occurred 15 km west to the CMF at depth of ~59 km [40].

Zone Z.17

In this zone, the most active Sagaing fault is present. Kundu and Gahalaut [41] suggested that earthquakes are generated by strike slip faulting on this Sagaing fault. It is responsible for the May 23, 1912 major earthquake ($M = 8$).

5 Catalogue Completeness

The Stepp test (1972) based on the statistical property of the Poisson distribution is used to obtain the time intervals during which the occurrence rate of earthquakes for the given range of magnitudes is uniform. This statistical test evaluates the stability of the mean rate of occurrences. If mean rate of occurrences is constant, it will show a straight line of slope $1/\sqrt{T}$ where standard deviation σ varies as $1/\sqrt{T}$. But if mean rate of occurrences is not uniform or not stable, σ deviates from the straight line of slope $1/\sqrt{T}$, where T is the time window in years. The time interval at which σ does not deviate from the straight line is defined as the completeness time interval for the given range of earthquake magnitude. For this purpose the number of earthquakes recorded during each decade (Since 1731-2015) for the given magnitude ranges were evaluated. Further, the completeness of the earthquake catalogue was also estimated using Cumulative Visual Inspection (CUVI) method. In this method, the cumulative number of earthquake events per year is plotted against the period of occurrence in years for each size of magnitude range. The completeness period based on both the methods was estimated for all identified seismic source zones and presented in figure form (Figs. 3 and 4) typically for one seismic source zone (Z.17) and tabulated for all source zones (Tables 1 and 2) and they are compared very well.

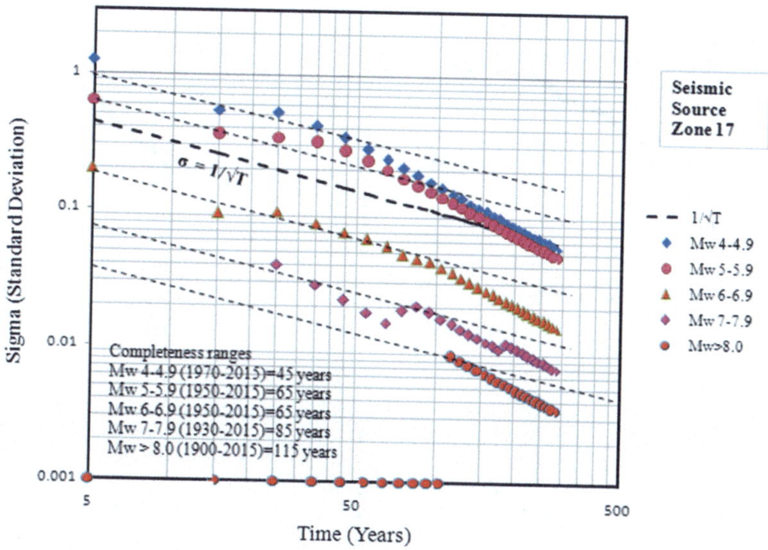

Fig. 3 The figure shows the completeness analysis of catalogue based on Stepp (1972) method

6 Methodology

6.1 Estimation of Seismicity Parameters

The seismic activity of a region is reported by seismicity parameters which correlate the magnitude of earthquake with its cumulative frequency of occurrences. '*a*' and '*b*' are the essential parameters in seismic hazard assessment. Therefore, systematic and precise estimation of seismicity parameters are required.

In this study, after superimposing of total 2508 seismic events (Mw $\geq$ 4) from prepared catalogue on the digitized tectonic map [Seismotectonic Atlas of India (SEISAT), 2000] number of events for each seismic source zone were counted. After data completeness analysis, out of 2508 events total 2461 events (Mw $\geq$ 4) were counted that occurred uniformly (with data completeness) in the study area. A total of 40 scattered events were outside the proposed source zones within the study area and hence were not taken into account for further study. Hence total 2421 events (Mw $\geq$ 4) were counted that occurred in the identified seismic source zones and observed maximum magnitude (Mw_{obs}) for each zone was also found (Table 3). After counting the events, '*a*' and '*b*' values were calculated for all seismic source zones by regression analysis (Table 3).

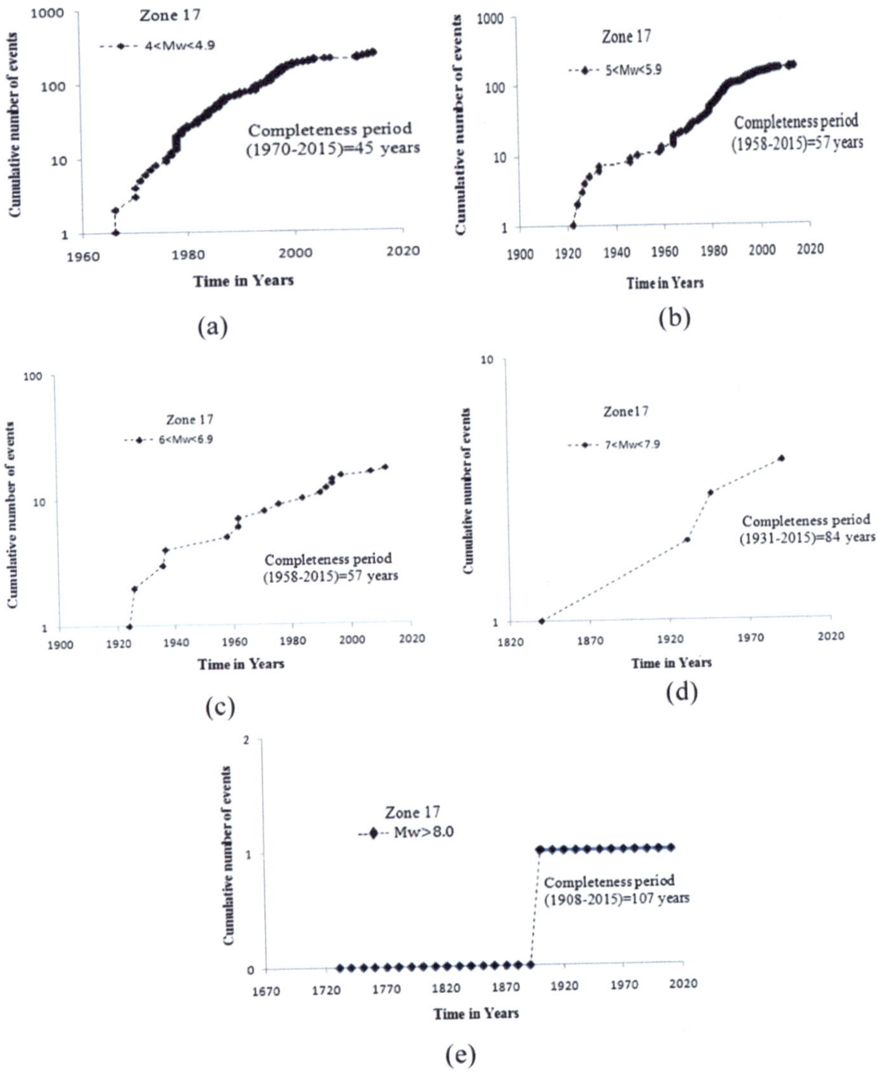

Fig. 4 a–e The figures show the completeness analysis of catalogue based on CUVI method

6.2 *Artificial Neural Network (ANN)*

A neural network is a computing system inspired by structural and functional mechanism of biological neural network. In ANN the network is formed by interconnected processing elements called neurons. A numeric weight associated with every link that connects the neurons describes the strength or amplitude of a connection between

Table 1 Completeness period (years) of catalogue based on Stepp (1972) method for different magnitude ranges (Mw)

Seismic source zones	4–4.9	5–5.9	6–6.9	7–7.9	>8.0
Z.1	1980–2015 = 35	1950–2015 = 65	1930–2015 = 85	1940–2015 = 75	No event
Z.2	1970–2015 = 45	1950–2015 = 65	1960–2015 = 55	1910–2015 = 105	1950–2015 = 65
Z.3	1990–2015 = 25	1960–2015 = 55	1820–2015 = 195	1930–2015 = 85	No event
Z.4	1980–2015 = 35	1970–2015 = 45	1930–2015 = 85	No event	1890–2015 = 125
Z.5	1990–2015 = 25	1980–2015 = 35	1930–2015 = 85	1940–2015 = 75	No event
Z.6	1990–2015 = 25	1970–2015 = 45	1940–2015 = 75	No event	No event
Z.7	1990–2015 = 25	1980–2015 = 35	1920–2015 = 95	No event	No event
Z.8	1990–2015 = 25	1980–2015 = 35	1960–2015 = 55	1840–2015 = 175	No event
Z.9	1990–2015 = 25	1970–2015 = 45	1940–2015 = 75	1860–2015 = 155	No event
Z.10	1980–2015 = 35	1960–2015 = 55	1940–2015 = 75	No event	No event
Z.11	1980–2015 = 35	1970–2015 = 45	1980–2015 = 35	1870–2015 = 145	No event
Z.12	1980–2015 = 35	1970–2015 = 45	1930–2015 = 85	1950–2015 = 65	No event
Z.13	1990–2015 = 25	1980–2015 = 35	1950–2015 = 65	No event	No event
Z.14	1980–2015 = 35	1970–2015 = 45	1920–2015 = 95	No event	No event
Z.15	1970–2015 = 45	1960–2015 = 55	1920–2015 = 95	1930–2015 = 85	No event
Z.16	1980–2015 = 35	1960–2015 = 55	1910–2015 = 105	1950–2015 = 65	No event
Z.17	1970–2015 = 45	1950–2015 = 65	1950–2015 = 65	1930–2015 = 85	1900–2015 = 115
Whole study area	1970–2015 = 45	1960–2015 = 55	1920–2015 = 95	1900–2015 = 115	1890–2015 = 125

Table 2 Completeness period (years) of catalogue based on CUVI method for different magnitude ranges (Mw)

Seismic source zones	4–4.9	5–5.9	6–6.9	7–7.9	>8.0
Z.1	1978–2015 = 37	1955–2015 = 60	1937–2015 = 78	1947–2015 = 68	No event
Z.2	1978–2015 = 37	1963–2015 = 52	1959–2015 = 56	1906–2015 = 109	1950–2015 = 65
Z.3	1992–2015 = 23	1966–2015 = 49	1822–2015 = 193	1930–2015 = 85	No event
Z.4	1979–2015 = 36	1966–2015 = 49	1933–2015 = 82	No event	1897–2015 = 118
Z.5	1992–2015 = 23	1976–2015 = 39	1932–2015 = 83	1943–2015 = 72	No event
Z.6	1991–2015 = 24	1966–2015 = 49	1948–2015 = 67	No event	No event
Z.7	1996–2015 = 19	1982–2015 = 33	1924–2015 = 91	No event	No event
Z.8	1991–2015 = 24	1980–2015 = 35	1963–2015 = 52	1842–2015 = 173	No event
Z.9	1998–2015 = 17	1979–2015 = 36	1956–2015 = 59	1868–2015 = 147	No event
Z.10	1987–2015 = 28	1959–2015 = 56	1950–2015 = 65	No event	No event
Z.11	1987–2015 = 28	1974–2015 = 41	1984–2015 = 31	1869–2015 = 146	No event
Z.12	1984–2015 = 31	1959–2015 = 56	1930–2015 = 85	1957–2015 = 58	No event
Z.13	1991–2015 = 24	1988–2015 = 27	1959–2015 = 56	No event	No event
Z.14	1985–2015 = 30	1973–2015 = 42	1920–2015 = 95	No event	No event
Z.15	1964–2015 = 51	1957–2015 = 58	1923–2015 = 92	1932–2015 = 83	No event
Z.16	1976–2015 = 39	1974–2015 = 41	1924–2015 = 91	1954–2015 = 61	No event
Z.17	1970–2015 = 45	1958–2015 = 57	1958–2015 = 57	1931–2015 = 84	1908–2015 = 107
Whole study area	1963–2015 = 52	1955–2015 = 60	1920–2015 = 95	1906–2015 = 109	1897–2015 = 118

Table 3 Seismicity parameters calculated for each zone

Seismic source zones	Seismicity parameters calculated for each zone				
	Seismicity Parameters		Coefficient of determination (R^2)	No. of main events/zone till 2015	Mw_{obs}
	a value	b value			
Z.1	4.69	0.84	0.93	333	7.7
Z.2	2.89	0.60	0.97	158	8.6
Z.3	2.54	0.60	0.98	53	7.2
Z.4	2.16	0.51	0.94	89	8.1
Z.5	2.70	0.61	0.98	61	7.3
Z.6	2.44	0.57	0.97	42	6.4
Z.7	2.38	0.58	0.96	44	6.8
Z.8	2.35	0.58	0.95	42	7.2
Z.9	1.79	0.50	0.99	29	7.6
Z.10	2.82	0.63	0.95	54	6.9
Z.11	2.81	0.65	0.97	82	7.2
Z.12	3.41	0.70	0.97	111	7.3
Z.13	2.96	0.71	0.93	26	6.0
Z.14	3.21	0.74	0.96	60	6.2
Z.15	4.15	0.73	0.98	510	7.4
Z.16	4.42	0.82	0.94	275	7.3
Z.17	3.91	0.70	0.98	452	8.2
Whole study area	4.98	0.77	0.98	2461	8.6

neurons. The output from each neuron is calculated by applying an activation function to the sum of inputs multiple by the weight vector. The ANN supports number of activation functions such as linear function, non-linear function, sign function, step function. In this study, we used a feedforward neural network with a backpropagation learning algorithm to train the network. A typical neural network in feedforward direction is given by:

$$O_j = f\left(\sum_{i=1}^{n}(w_{ij}\, x_i) + b_j\right) \tag{6}$$

where x_i is the input vector, O_j is the output vector, w_{ij} is a weight factor between two neurons, b_j is the bias weight vector. In this study, input layer neurons are linear whereas neurons in the hidden and output layers have non-linear and differentiable Log-Sigmoid signal function. This function is given by:

$$f(y) = \frac{1}{1 + e^{-y}} \tag{7}$$

The backpropagation learning algorithm is based on a generalized delta-rule accelerated by a momentum term. To improve training rate the changes in the weight factor are accelerated by introducing momentum term. The weight factors and bias are adjusted by using Eqs. (8) and (9).

$$\left(w_{ij}^{k+1}\right) = \left(w_{ij}^{k}\right) + \eta\left(\delta_j^k\right)O_i + \alpha * \Delta\left(w_{ij}^{k-1}\right) \quad (8)$$

$$\left(b_j^{k+1}\right) = \left(b_j^{k}\right) + \eta\left(\delta_j^k\right) + \alpha * \Delta\left(b_j^{k-1}\right) \quad (9)$$

where 'η' is the learning rate, 'α' is the momentum coefficient, 'w_{ij}' is the weight factor associated between two neurons 'b_j' is the biass weight, 'O' is the output 'δ' is the gradient descent correction term and 'k' stands for number of patterns.

6.2.1 Input Parameters

To train the neural network total five input parameters namely longitude, latitude, elapsed time between two consecutive earthquake events, cumulative magnitude probability and seismic energy were used.

- **Location and elapsed time:** Longitudes, latitudes and elapsed time between two events of past seismicity in the region were taken from the latest earthquake catalogue which was prepared in this study.
- **Cumulative magnitude probability:** The CDF for the Gutenberg-Richter law with upper and lower bounds can be expressed as cumulative magnitude probability F_M (m) which is governed by *b*-value and it is given as:

$$F_M(m) = \frac{10^{a-bm} - 10^{a-bm_o}}{10^{a-bm_{max}} - 10^{a-bm_o}} = \frac{10^{-bm} - 10^{-bm_o}}{10^{-bm_{max}} - 10^{-bm_o}} \quad (10)$$

  where, m is the magnitude greater than or equal to threshold magnitude m_o but less than or equal to maximum magnitude m_{max}. 'a' and 'b' values are the seismicity parameters which are estimated in this study for all 17 seismic source zones. The maximum magnitude m_{max} has been calculated using incremental value method [42]. In this method to estimate the m_{max} an increment of 0.5 is added to the observed maximum magnitude (Mw_{obs}) [17].
- **Seismic energy:** Gutenberg and Richter [43] proposed one equation to estimate the total seismic energy during the earthquake which is governed by the magnitude of earthquake and it is given as:

$$\log_{10}(E) = 11.8 + 1.5M_s \quad (11)$$

  where E is expressed in ergs. This equation is also applicable for moment magnitude.

6.2.2 Performance Evaluation of ANN

The performance of ANN in predicting the earthquake magnitude is evaluated by estimating the mean absolute error, sum of squared error and mean squared error.

- **Mean absolute error (MAE):** It is the measure of accuracy that how closer predicted values are to the target values and it is given as:

$$MAE = \frac{1}{k}\sum_{i=1}^{k}|T_i - O_i| \tag{12}$$

- **Sum of squared error (SSE):** This quantity measures the sum of the squares of the errors between predicted and target values and it is given as:

$$SSE = \sum_{i=1}^{k}(T_i - O_i)^2 \tag{13}$$

- **Mean squared error (MSE):** This quantity measures the average of the squares of the errors between predicted and target values and it is given as:

$$MSE = \frac{1}{k}\sum_{i=1}^{k}(T_i - O_i)^2 \tag{14}$$

where, T = target value of output variables, O = predicted value of output variables, k = total number of training data sets.

6.2.3 Training and Testing of Neural Network

Smaller magnitude earthquakes do not cause significant damage in the buildings or in other structures. Therefore, in predicting the earthquake magnitude, the events of $Mw \geq 6.0$ have been considered for further study. Hence, it was noticed that out of 17 identified seismic source zones only 5 zones are seismically very active and produce earthquakes of $Mw \geq 6.0$ frequently and they are Z.1, Z.2, Z.15, Z.16, Z.17 (Fig. 5). Therefore, only 5 seismic source zones were used for further study where reasonable amounts of main shock events of magnitude $Mw \geq 6.0$ have been found from the year 1731 to 2015 in the study area. The architecture of a typical neural network with five input parameters (longitude, latitude, elapsed time between two consecutive earthquake events, cumulative magnitude probability and seismic energy) and one output parameter (magnitude of earthquake) is shown in Fig. 6.

Further, input parameters were calculated for all five seismic source zones and presented typically for one seismic source zone (Z.17) in Table 4. The neural networks were trained using estimated input and output parameters for the earthquake events occurred from year 1731 to 1990. The trained networks were tested using input

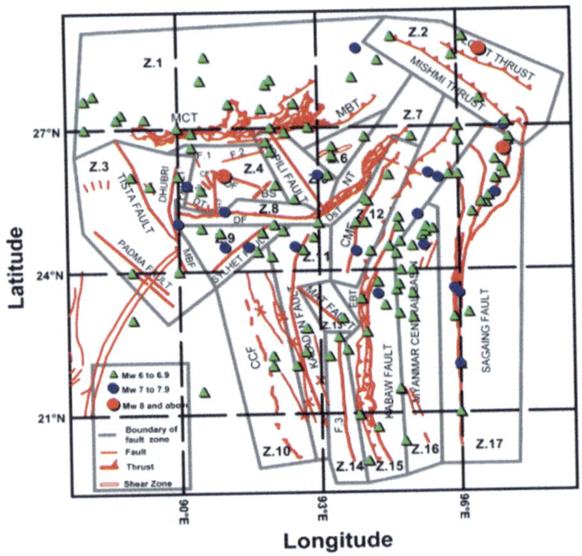

Fig. 5 Map shows epicentral distribution of earthquakes (Mw ≥ 6.0) used for prediction of earthquake magnitudes in the study area collected from year 1731–2015

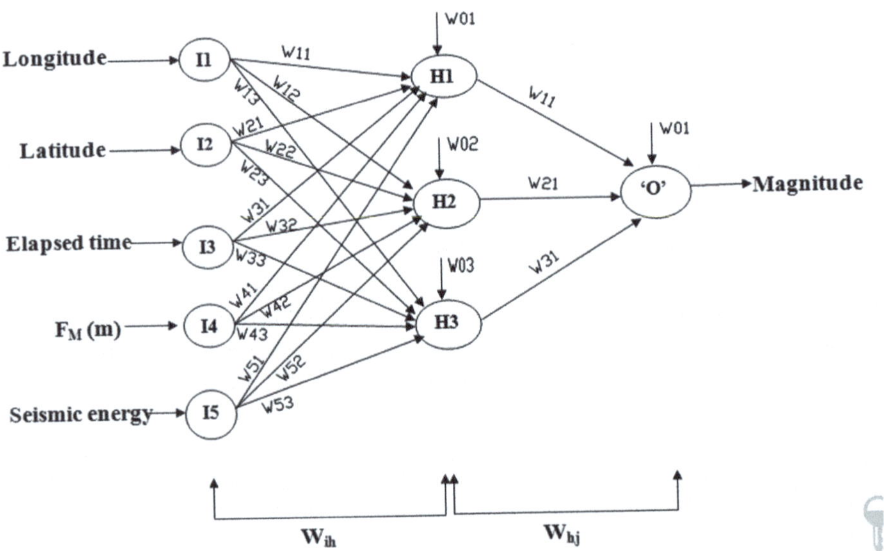

Fig. 6 Architecture of an artificial neural network with weight index ranges between input and hidden layer i = 0, 1, 2, 3, 4, 5; h = 1, 2, 3 and between hidden and output layer h = 0, 1, 2, 3; j – 1

Table 4 Earthquake events (Mw ≥ 6.0) occurring since 1731–2015 in the identified seismic source zone

Zone No.	Longitude (°)	Latitude (°)	Date	Time (hh:mm)	Time elapsed (years)	F_M (m)	Seismic energy (ergs)	Mw
17	96.00	22.00	23/03/1839	06:25	00.00	0.84	2.82E+22	7.1
	97.00	26.50	12/12/1908	12:54	69.72	0.98	1.26E+24	8.2
	96.00	26.00	01/8/1924	14:42	15.64	0.28	1.26E+21	6.2
	97.00	26.00	10/05/1926	08:19	01.77	0.39	1.78E+21	6.3
	96.80	25.40	27/01/1931	20:09	04.71	0.94	1.58E+23	7.6
	96.90	26.40	19/06/1936	16:34	05.39	0.28	1.26E+21	6.2
	96.80	25.90	31/08/1937	14:15	01.20	0.28	1.26E+21	6.2
	96.20	23.90	12/09/1946	15:17	09.03	0.92	1.12E+23	7.5
	96.27	25.22	28/10/1958	05:22	12.13	0.28	1.26E+21	6.2
	96.94	26.13	20/02/1962	22:02	03.31	0.56	3.55E+21	6.5
	96.91	26.49	22/09/1962	06:51	00.59	0.48	2.51E+21	6.4
	96.51	25.26	30/05/1971	06:25	08.69	0.39	1.78E+21	6.3
	97.06	26.69	12/08/1976	06:25	05.20	0.56	3.55E+21	6.5
	97.09	26.70	28/11/1984	06:25	08.29	0.15	8.91E+20	6.1
	96.20	23.10	09/01/1990	06:25	05.11	0.69	7.08E+21	6.7
	95.95	23.59	05/01/1991	14:57	00.99	0.91	7.94E+22	7.4
	95.96	24.01	15/06/1992	06:25	01.44	0.73	1.00E+22	6.8
	96.00	21.00	29/05/1994	06:25	01.96	0.28	1.26E+21	6.2
	96.71	25.54	21/11/1994	06:25	00.48	0.63	5.01E+21	6.6
	96.61	25.37	30/12/1997	06:25	03.11	0.00	6.31E+20	6.0
	96.60	25.47	29/06/2007	06:25	09.50	0.15	8.91E+20	6.1
	95.89	23.01	11/11/2012	01:12	05.37	0.73	1.00E+22	6.8

parameters for the earthquake events occurred from year 1991 to 2015. The predicted magnitudes of ANN in the years 1991 to 2015 were compared with originally seismometer recorded magnitudes. And then MAE, SSE, and MSE were estimated for all 5 considered zones.

Before starting the training, data of the patterns were normalized to be less than 1.0. For example, in Tables 4 and 5, estimated input and output parameters and normalized input and output parameters, training of network, updated weights and errors are shown respectively, typically for one source zone (Z.17). In Table 5, 'T' is the target output (magnitude of earthquakes, normalized by factor 10) and I1, I2, I3, I4 and I5 (longitude, latitude, elapsed time between two consecutive earthquake events, cumulative magnitude probability and seismic energy, normalized by 100, 30, 80, 1.0 and $1.5*10^{24}$ respectively) are input parameters. First the neural networks were trained for different values of learning rate (η) and momentum coefficient (α).

Table 5 Training of an artificial neural network for events occurred from year 1731–1990 (No. of epochs 5000)

Source zone	Input patterns					"T"	K	Weights				"O"	Error = T – O
	I1	I2	I3	I4	I5			Wih		Whj			
Z.17	0.970	0.883	0.872	0.984	0.839	0.820	14	W01=	0.457	W01=	1.032	0.821	–0.001
	0.960	0.867	0.195	0.279	0.001	0.620		W02=	–0.442	W11=	0.651	0.621	–0.001
	0.970	0.867	0.022	0.389	0.001	0.630		W03=	1.870	W21=	1.268	0.625	0.005
	0.968	0.847	0.059	0.936	0.106	0.760		W11=	0.709	W31=	–2.029	0.751	0.009
	0.969	0.880	0.067	0.279	0.001	0.620		W12=	0.142			0.616	0.004
	0.968	0.863	0.015	0.279	0.001	0.620		W13=	0.822			0.613	0.007
	0.962	0.797	0.113	0.923	0.075	0.750		W21=	–0.478			0.746	0.004
	0.963	0.841	0.152	0.279	0.001	0.620		W22=	1.283			0.618	0.002
	0.969	0.871	0.041	0.561	0.002	0.650		W23=	0.919			0.652	–0.002
	0.969	0.883	0.007	0.482	0.002	0.640		W31=	0.362			0.638	0.002
	0.965	0.842	0.109	0.389	0.001	0.630		W32=	0.581			0.628	0.002
	0.971	0.890	0.065	0.561	0.002	0.650		W33=	0.255			0.653	–0.003
	0.971	0.890	0.104	0.151	0.001	0.610		W41=	0.590			0.608	0.002
	0.962	0.770	0.064	0.686	0.005	0.670		W42=	–0.300			0.677	–0.007
								W43=	–3.066				
								W51=	–0.020				
								W52=	1.594				
								W53=	–0.352				
Mean absolute error=			0.0035										
Sum of squared error=			0.0002										
Mean squared error=			0.00002										

It was observed that for $\eta = 0.7$ and $\alpha = 0.9$ network yields stable solution and it reaches to desired output value very fast. The updated weights after 5000 training epochs assigned to the neural networks are shown in Table 5. W_{ih} is the weight assigned between input layer and hidden layer, W_{hj} is the weight assigned between hidden layer and output layer and "O" is the predicted output. Similarly for other 4 seismic source zones networks were trained using data till the year 1990, the trained networks were used to predict the magnitude of earthquakes occurred in the years 1991 to 2015 (Table 6).

7 Results and Discussion

The highest value of '*a*' as 4.69 was observed in Z.1 which includes the MBT and MCT in the study area and for zones Z.14 to Z.17 '*a*' value varies from 3.21 to 4.42 which include the Indo Burmese Wedge and Sagaing fault region. In the other zones '*a*' value varies from 1.79 to 3.41, these zones are seismically less active as compared to zones Z.1, Z.14 to Z.17 but may be capable in producing the higher magnitude earthquakes.

The lowest value of '*b*' was observed in Z.9 as 0.50. This zone has experienced the 1918 Srimangal earthquake. The second lowest value of '*b*' was observed in Z.4 as 0.51. This zone has experienced the 1897, Shillong earthquake. It was observed that in zones Z.2 to Z.9 the '*a*' value varies from 1.79 to 2.89 and '*b*' value varies from 0.5 to 0.61 which indicates the less seismic activity of the zones as compared to zones Z.1, Z.14 to Z.17 but it also indicates the lower proportion of smaller earthquakes to large earthquakes and capability in producing the higher magnitude earthquakes in such regions. The Z.2 has experienced the great Assam earthquake in 1950. In the Indo Burmese Wedge and Sagaing fault zones the estimated '*b*' value normally varies from 0.70 to 0.82 as presented in Table 3.

The present results were validated with the results derived by the other researchers. Although their earthquake catalogue and methodology are different, it still would be comparable with the present results. Raghukanth [44] estimated seismicity parameters for India by dividing the whole area into grids of size $1°*1°$ along E-W and N-S directions. He suggested that in the north-eastern region of India, '*b*' value varies from 0.7 to 0.85. In the present study we estimated '*b*' value for the whole study area using catalogue from year 1731 to 2015 and found '*b*' value as 0.77 (Table 3) which is existing in between the range suggested by Raghukanth [44]. However, Das et al. [45] had also divided the NE India into nine seismogenic zones and found magnitude of completeness and seismicity parameters for each source zone. The '*b*' value calculated in their study and estimated '*b*' value in the present study are found to be very close and comparable.

In this study it was noticed that out of 17 identified seismic source zones, only five zones are seismically very active and produce earthquakes ($Mw \geq 6.0$) frequently. The trained neural networks considering earthquake catalogue since 1731 to 1990 were used to predict the magnitude of earthquakes occurred in the years 1991 to

Table 6 Performance evaluation of ANN in predicting the magnitude of earthquakes occurred from year 1991 to 2015 in Northeast India

Seismic source zone	Events date occurred since 1991–2015	ANN predicted magnitude of earthquakes occurred between 1991 and 2015	Actual magnitude of earthquakes occurred between 1991 and 2015	Success rating in %	Errors
Z.1	17/02/1995	6.187	6.2	99.79	MAE = 0.042
	26/09/1998	6.318	6.3	100.29	SSE = 0.009
	18/09/2011	6.806	6.9	98.64	MSE = 0.003
Z.2	24/05/1993	6.066	6.1	99.44	MAE = 0.051
	07/06/2000	6.732	6.8	99.00	SSE = 0.006
					MSE = 0.003
Z.15	20/12/1991	6.096	6.1	99.93	MAE = 0.003
	27/03/1992	5.962	6.0	99.37	SSE = 0.009
	29/05/1994	6.652	6.6	100.79	MSE = 0.001
	07/07/2002	6.083	6.1	99.72	
	15/02/2005	5.938	6.0	98.97	
	11/05/2006	6.794	6.8	99.91	
	04/02/2011	6.220	6.2	100.32	
Z.16	05/01/1991	6.850	6.9	99.28	MAE = 0.136
	08/08/1994	6.276	6.3	99.62	SSE = 0.287
	06/05/1995	6.838	6.9	99.10	MSE = 0.032
	19/11/1996	5.936	6.1	97.31	
	11/10/2000	5.707	6.0	95.12	
	21/09/2003	6.625	6.6	100.38	
	18/09/2005	6.272	6.3	99.56	
	27/07/2008	5.705	6.0	95.08	
	21/09/2009	5.719	6.0	95.32	
Z.17	05/01/1991	7.381	7.4	99.74	MAE = 0.061
	15/06/1992	6.874	6.8	101.09	SSE = 0.036
	29/05/1994	6.066	6.2	97.84	MSE = 0.005
	21/11/1994	6.630	6.6	100.45	
	30/12/1997	5.947	6.0	99.12	
	29/06/2007	6.068	6.1	99.48	
	11/11/2012	6.888	6.8	101.29	

2015 in the considered 5 zones. The success rating for the predicted magnitudes of the proposed ANN model varies from 95% to 101% in the different seismic source zones.

8 Conclusion

The northeast region of India is seismically most active region in the world. The seismicity parameters estimated in study area can be used in microzonation projects and hazard assessment by probabilistic seismic hazard analysis (PSHA). And the ANN model presented in this study yields higher prediction accuracy for magnitude of earthquakes in the northeast India because of its capability of capturing non-linear relationship. The main advantage of ANN is that no mathematical relationship is required between input and output parameters. The proposed ANN model would be expected to help in the domain of earthquake forecasting in the northeast region of India.

Acknowledgements The authors acknowledge the support of SERB-DST (ECR/2016/001329) to carry out the research studies.

References

1. B. Gutenberg, C.F. Richter, Frequency of earthquakes in California. Bull. Seismol. Soc. Am. **34**, 185–188 (1944)
2. M. Wyss, K. Shimazaki, S. Wiemer, Mapping active magma chambers by b values beneath the off-Ito volcano. J. Geophys. Res. **102**(20), 413–420 (1997)
3. S. Weimer, M. Wyss, Mapping the frequency magnitude distribution in asperities; an improved technique to calculate recurrence times? J. Geophys. Res. **102**(15), 115–128 (1997)
4. C. Frohlich, S.D. Davis, Teleseismic b values; or, much ado about 1.0. J. Geophys. Res. **98**, 631–644 (1993)
5. K. Mogi, Fracture of rocks. Bull. Volcanol. Soc. Jpn. **7**, 89–101 (1962)
6. N.W. Warren, G.V. Latham, An experimental study of thermally induced microfracturing and its relation to volcanic seismicity. J. Geophys. Res. **75**, 4455–4464 (1970)
7. F.S. Wong, A.T.Y. Tung, W. Dong, Seismic hazard prediction using neural nets, in *Earthquake Engineering, Tenth world Conference*, Balkema, Rotterdam (1992), pp. 339–343
8. A.S.N. Alarifi, N.S.N. Alarifi, S.A. Humidan, Earthquake magnitude prediction using artificial neural network in northern red sea area. J. King Saud Univ. Sci. **24**, 301–313 (2012)
9. M. Moustra, M. Avraamides, C. Christodoulou, Artificial neural networks for earthquake prediction using time series magnitude data or seismic electric signals. Expert Syst. Appl. Int. J. **38**(12), 15032–15039 (2011)
10. S. Narayanakumar, K. Raja, A BP artificial neural network model for earthquake magnitude prediction in Himalayas, India. Circuits Syst. **7**, 3456–3468 (2016)
11. P. Molnar, P. Tapponnier, Cenozoic tectonics of Asia: effects of a continental collision. Science **189**, 419–426 (1975)
12. J.R. Kayal, Seismicity of northeast India and surroundings, Development over the past 100 years. J. Geophys. Res. **19**(1), 9–34 (1998)

13. R. Hall, Cenozoic plate tectonic reconstructions of SE Asia, in *Petroleum Geology of Southeast Asia*, vol. 126, ed. by A.J. Fraser, S.J. Methews, R.W. Murphy (Geological Society of London Special Publication, 1997), pp. 11–23
14. B. Olympa, A. Kumar, A review on the tectonics setting and seismic activity of the Shillong Plateau in the light of past studies. Disaster Adv. **8**(7), 34–45 (2015)
15. Bureau of Indian Standards, *IS 1893 (Part I): Indian Standard Criteria for Earthquake Resistant Design of Structures* (Bureau of Indian Standards, New Delhi, India, 2002)
16. A. Sil, T.G. Sitharam, S. Kolathayar, Probabilistic seismic hazard analysis of Tripura and Mizoram states. Nat. Hazards **68**(2), 1089–1108 (2013)
17. T.G. Sitharam, A. Sil, Comprehensive seismic hazard assessment of Tripura & Mizoram states. J. Earth Syst. Sci. **123**(4), 837–857 (2014)
18. J.K. Gardner, L. Knopoff, Is the sequence of earthquakes in southern California with aftershocks removed, Poissonian? Bull. Seismol. Soc. Am. **64**(5), 1363–1367 (1974)
19. R.A. Uhrhammer, Characteristics of northern and central California seismicity. Earthq. Notes **1**, 21 (1986)
20. J.R. Kayal, Microearthquake activity in some parts of the Himalaya and the tectonic model. Tectonophysics **339**, 331–351 (2001)
21. P. Molnar, A review of the seismicity and the rates of active underthrusting and deformation at the Himalaya. J. Himalayan Geol. **1**, 131–154 (1990)
22. L. Seeber, J. Armbruster, Great detachment earthquakes along the Himalayan Arc and long-term forecasting, in: *Earthquake Prediction: An International Review, Maurice Ewing Series*, vol. 4, ed. by D.W. Simpson, P.G. Richards (American Geophysical Union, 1981), pp. 259–277
23. R. Bilham, P. England, Plateau pop-up during the 1897 Assam earthquake. Nature **410**, 806–809 (2001)
24. G.K. Nayak, V.K. Rao, H.V. Rambabu, J.R. Kayal, Pop-up tectonics of the Shillong plateau in the great 1897 earthquake (Ms 8.7), Insight from the gravity in conjunction with the recent seismological results. Tectonics **27**, TC 1018 (2008). https://doi.org/10.1029/2006tc002027
25. D.K. Saha, D.C. Naskar, P.M. Bhattacharya, J.R. Kayal, Geophysical and seismological investigations for the hidden Oldham fault in the Shillong plateau and Assam valley of northeast India. J. Geol. Soc. India **69**, 359–372 (2007)
26. C.P. Rajendran, K. Rajendran, B.P. Duarah, S. Baruah, A. Earnest, Interpreting the style of faulting and paleoseismicity associated with the 1897 Shillong, northeast India, earthquake: Implications for regional tectonism. Tectonics. **23**, TC4009 (2004). https://doi.org/10.1029/2003tc001605
27. S. Baruah, D. Hazarika, A. Kalita, S. Goswami, Intrinsic and scattering attenuation in Chedrang fault and its vicinity—the rupture area of great Assam earthquake of 12 June 1897 (M = 8.7). Curr. Sci. **99**(6), 775–784 (2010)
28. J.R. Kayal, S.S. Arefiev, S. Baruah, D. Hazarika, N. Gogoi, A. Kumar, S.N. Chowdhury, S. Kalita, Shillong plateau earthquakes in northeast India region: complex tectonic model. Curr. Sci. **91**, 109–114 (2006)
29. P.M. Bhattacharya, J.R. Kayal, S. Baruah, S.S. Arefiev, Earthquake source zones in northeast India: seismic tomography, fractal dimension and b-value mapping. Pure Appl. Geophys. **167**(8), 999–1012 (2010)
30. D.R. Nandy, *Geodynamics of Northeastern India and the adjoining region* (ACB Publications, Kolkata, 2001)
31. R.D. Oldham, The Cachar earthquake of 10th January, 1869. Mem. Geol. Soc. India **19**, 1–98 (1882)
32. R.K.J. Singh, *A Short History of Manipur* (Manipur Sahitya Parishad, Imphal, India, 1965), p. 365
33. K.N. Khattri, M. Wyss, Precursory variation of seismic rate in the Assam area, India. Geology **6**, 685–688 (1978)
34. J.R. Kayal, Earthquake source process in northeast India: a review. Himalayan Geol. **17**, 53–69 (1996)

35. H.M. Chaudhury, H.N. Srivastava, Seismicity and focal mechanism of some recent earthquakes in northeast India and neighbourhood. Ann. Geophys. **29**(1–2), 41–56 (1976)
36. M. Morino, A.S.M.M. Kamal, D. Muslim, R.M.E. Ali, M.A. Kamal, M.Z. Rahman, F. Kaneko, Seismic event of the dauki fault in 16th century confirmed by trench investigation at Gabrakhari village, Haluaghat, Mymensingh, Bangladesh. J. Asian Earth Sci. **42**, 492–498 (2011)
37. T. Maurin, C. Rangin, Structure and kinematics of the Indo-Burmese Wedge: recent and fast growth of the outer wedge. Tectonics **28**, 1–33 (2009)
38. A.M. Sikder, M.M. Alam, 2-D modelling of the anticlinal structures and structural development of the eastern fold belt of the Bengal Basin, Bangladesh. Sediment. Geol. **155**, 209–226 (2003)
39. A. Kumar, M. Sanoujam, L. Sunil, T. Dolendro, Active deformations at the Churachandpur Mao fault (CMF) in Indo Burma ranges: Multidisciplinary evidences. Int. J. Geosci. **2**, 597–609 (2011)
40. V.K. Gahalaut, B. Kundu, The 4 January 2016 Manipur earthquake in the Indo Burmese wedge, an intra-slab event. Geomatics Nat. Hazards Risk **7**(5), 1506–1512 (2016). https://doi.org/10.1080/19475705.2016.1179686
41. B. Kundu, V.K. Gahalaut, Earthquake occurrence processes in the Indo-Burmese wedge and Sagaing fault region. Tectonophysics **524**, 135–146 (2012)
42. I.D. Gupta, The state-of-the-art in seismic hazard analysis. ISET J. Earthq. Technol. **39**(4), 311–346 (2002)
43. B. Gutenberg, C. Richter, Earthquake, magnitude, intensity, energy and acceleration. Bull. Seismol. Soc. Am. **46**, 105–145 (1956)
44. S.T.G. Raghukanth, Estimation of seismicity parameters for India. Seismol. Res. Lett. **81**(2), 207–217 (2010)
45. R. Das, H.R. Wason, M.L. Sharma, Temporal and spatial variations in the magnitude of completeness for homogenized moment magnitude catalogue for northeast India. J. Earth Syst. Sci. **121**, 19–28 (2012)

A New Approach for Assessment of the Coating Mortar Adherence in Ancient Masonries Through Ultrasonic Data

Emanuel Araújo, Maycon Bessa, Israel Sousa, Amanda Fontenele, Rosineide Luz, Mylene Vieira, and Esequiel Mesquita

Abstract In this work ultrasonic pulse velocity (UPV) was used to assess the coating mortar adherence of clay bricks masonries. As a contribution to employment of non-destructive tests to historic buildings characterization, this paper has as the main aim to introduces a new methodology for damage identification (especially the presence of cracks) and coating mortar detachments in masonry walls. For that, an experimental setup based on ultrasonic measurements were carried out on the external walls of the Nossa Senhora da Expectação Church, built in 1709, at Icó, Brazil. The results showed that ultrasonic pulse velocity mapping can be used to identify the adherence and the non-adherence between coating mortar and the masonry. Also, a standard pattern of ultrasonic velocity distribution for masonry was identified. Finally, in this paper a new point of view on UPV interpretation was introduced, namely based on analysis the pattern of waves distribution.

Keywords Ultrasonic pulse velocity · NDT · Coating mortar · Adherence · Historic buildings

1 Introduction

The development of non-destructive strategies and methodologies to support the maintenance, retrofitting and structural safe of historic buildings has been a topic of increasing interest by technical community, especially since from 60 years. Especially concerning masonry structures due to the damage mechanisms, constructive issues and materials heterogeneity, to assess masonries by non-destructives tests (NDT) has been a complex task. While since from 60 decade several advances have been done in isotropic materials, as steel and concrete, applications of NDT on anisotropic materials as masonry are rarely reported in the literature [1].

E. Araújo · M. Bessa · I. Sousa · A. Fontenele · R. Luz · M. Vieira · E. Mesquita (✉)
LAREB, Department of Civil Engineering, Federal University of Ceará, Campus Russas, Ceará 62900-000, Brazil
e-mail: emesquita@ufc.br

J. M. P. Q. Delgado (ed.), *Case Studies in Building Constructions*, Building Pathology and Rehabilitation 15, https://doi.org/10.1007/978-3-030-55893-2_4

NDT can be a very useful tool for analysis of the structural characteristics of historic masonries. Through correlation between sonic waves velocities and mechanical properties from samples experimentally tested in laboratory, it is possible estimating mechanical parameters of the structure in situ [2]. Nonetheless, in this case, the results are limited to direct ultrasonic measurements. But, even with these limitations, ultrasonic tests stand out as quite promising NDT.

In addition, ultrasonic data has been extensively used for masonries characterization, as the works developed by [3, 4], but also limited on interpretation of ultrasonic velocities pulse. Although promising, the use of more sophisticated statistical methods in the analysis of ultrasonic data to assess the condition of a building has not been widely explored, as well the interpretation of the pattern distribution of ultrasonic waves over an area are not commonly reported on the literature.

Mesquita et al. [5] and Kurata et al. [2] highlight that despite retrofitting measures in historic buildings has been performed in the last years, less intrusive techniques should be employed to preserve the historical characteristics and new damage introduction. Heritage constructions can be assigned as a special case, because they present structural systems different from the current ones, with complex behavior that often cannot be assessed or understood through the current codes, standards, techniques, or devices.

According to ICOMOS [6], information regarding the constructive characteristics and materials are insufficient to decide the adequate retrofitting technique. It is needed to assess the whole context of the building which may include materials, systems, cultural, physical and historical conditions to define a more appropriate methodologies and intervention methods to be employed.

Among the others existing NDT, the ultrasonic pulse velocity (UPV) should be highlighted. The method consists in measures the time of propagation of the ultrasonic wave through a material. The method involves the measurement of the distance of the propagation to assess the speed of sound and correlate with the quality and type of material tested.

In the case of masonry, Binda et al. [7] report that the UPV tests are useful to evaluate parameters such as the quality of masonry through the morphology of the wall section, the presence of voids and flaws, the cracks and patterns; and the detection of changes on the physical characteristics of the material. However, the characterization of masonry is a hard task due to its typological variability and anisotropic properties [8].

Recent advances on employment of ultrasonic techniques for masonry characterization have point to:

- Sonic and ultrasonic tests can be used to assess the effect of retrofitting techniques in masonries, as showed by [9];
- Sonic and ultrasonic tests can used in a combined way with others NDT and semi-destructives strategies for estimating anisotropic material mechanical properties, as presented by [7, 10];
- The masonry joints can influence the waves behavior of sonic and ultrasonic tests [11, 12];

- Ultrasonic tests can be used to identify heterogeneities in ancient masonries and to identify if a join of walls presents similar constructive properties, as well the work also shows that the loading and opening sections can influence the ultrasonic waves behavior in masonries [3];
- The presence of a mortar recovery layer (15 mm of thickness) does not presents significant influences on the ultrasonic wave behavior, as suggested by [8, 14].

However, gaps still extant in the understanding of characterization of the historic masonries. Thus, some questions regarding the use of indirect UPV method on the characterization of historical masonry walls should be investigated, such as:

1. Does the UPV test be applied to assessment of coating layer of masonry?
2. How does the presence of voids, flaws, imperfections and changes in temperature and humidity should influence the propagation of ultrasonic waves?
3. Does the UPV technique obtain reliable mechanical parameters of masonry walls?

Aiming to contribute to overcoming the above-mentioned issues, this paper has as the main aim to introduces a new methodology for identify damages, presence of cracks, and specially coating mortar detachments in masonry walls. For that the experimental setup was carried out on the external walls of the Nossa Senhora da Expectação Church, with the use of UPV indirect technique. Built in 1709, the Church is considered the older construction of Icó city, Ceará—Brazil. Furthermore, the study aims to contribute to the enlargement of the ultrasonic method application on the characterization of historical heritage.

2 Nossa Senhora Da Expectação Church

Nossa Senhora da Expectação Church was built-in 1709. The Church is considered the older construction of Icó city, Ceará, Brazil and one of the examples of vernacular Baroque. A unique sample of the Luso-Brazilian architecture.

Over the years several interventions in the structure of the Church were performed in the original, changing the original architecture. Information on all changes are not easily found, especially due to the absence of registers. However, the most recent interventions were identified:

- In 1999, the original lime-based-paint of the walls and the wood frames in colonial green of the Church were restored. The stairs of the choir's room and the wood structure of the roof were also rehabilitated;
- In 2004, a new intervention was performed in the pillars due to infiltrations caused by the high content of saltpeter in the soil composition. Besides, the external walls and the main door were painted, the internal lightening system was changed.

The Church has one tower on the left side (Fig. 1) with 18.07 m of height, a base of octagonal shape and a campanile on its top.

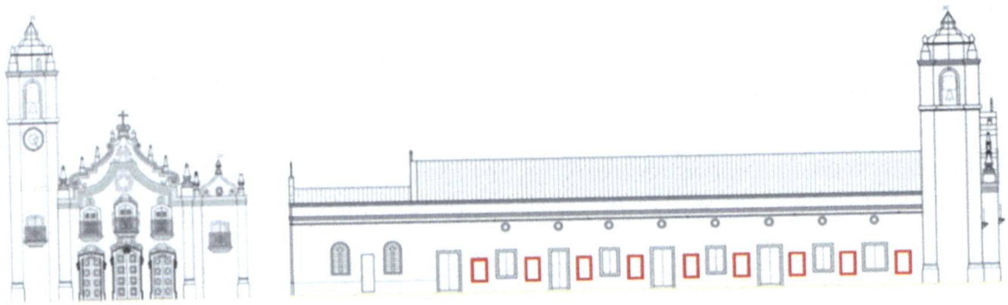

Fig. 1 Frontal façade and lateral section of the Nossa Senhora da Expectação Church

The Nossa Senhora da Expectação Church is composed by a main nave, two lateral naves, separated each one by four arch opening from the main nave, and a set of doors and windows with external access. Geometrically, the Church comprises a rectangular structure, measuring 45.47 m along of the longitudinal direction and 16.64 m along of the transversal direction. Internally, ten distinct spaces such as, the main nave (A), lateral naves (B) and (C), Altar-Mor (D), hall (E) and (F), sacristy (G), secretary room (H), Chapel (I) and deposit (J), comprise the building, as detailed in Fig. 2.

20 panels of external lateral walls namely identified as P01 to P10 at the lateral northeast of the Church and as P11 to P20 at the lateral southeast of the Church (see Fig. 2) presenting similar configuration and geometry were selected for the study.

The structure of the Church, built with the use of vernacular techniques is composed of clay bricks masonry with thickness varying between 46 and 90 cm. The clay bricks with an average dimension of 50 × 25 × 5 cm were joined with a layer of lime mortar varying between 1.00 and 1.50 cm of thickness.

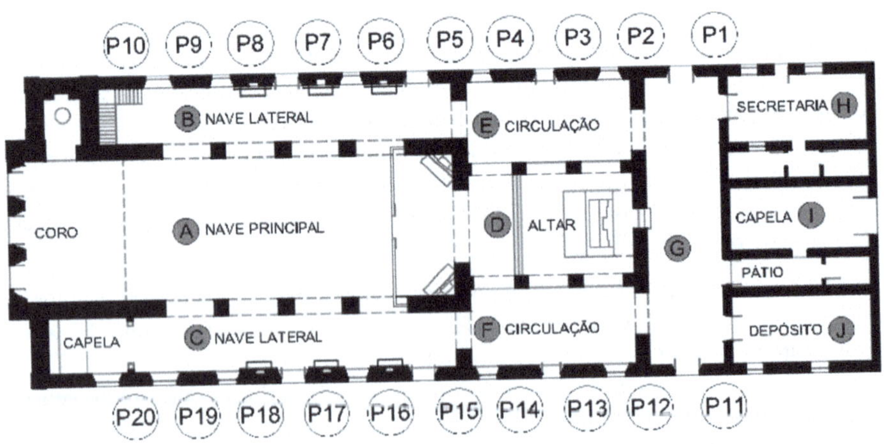

Fig. 2 Plant of the Nossa Senhora da Expectação Church

3 Experimental

For this study, the ultrasonic velocities were analyzed based on the time of its propagation across panels of an ancient masonry structures: a set of solid clay brick blocks laid with lime mortar and covering with the lime mortar too, with an average thickness of 1.5 cm. As the ultrasonic waves presents different velocities when propagating by solid, liquid, and even gaseous environment. Also, for each one material it is possible to identify a typical range of ultrasonic propagation. So, based on that it is possible to use the average speed of propagation to analyze the different properties of the materials.

The basic idea is to pass the ultrasonic wave inside the material, measuring the time of its propagation to another point. Once the distance between the points from which the wave was transmitted and received was known, it is then possible to determine the average speed in the propagation section.

In this experimental setup Proceq® PUNDIT 2000® 54 kHz equipment was used. Data processing was performed based on the recommendations of NBR 16805 [13]. The indirect ultrasonic pulse transmission method consists in the positioning of the transducer-transmitter (T) and transducer-receiver (R) on the same surface of the analyzed material, as can be seen by Fig. 3.

Initially the Church walls were divided in twenty panels. The panels were identified from P1 to P20. The panels P1 to P10 are located in the Northeast façade of the Church, while panels from P11 to P20 are located in the Southwest lateral. During the performance of the experimental procedure, only 16 of the 20 panels were actually tested, since some of them were intercommunicated with other structures.

Ultrasonic tests were performed for each wall considering the horizontal direction (direction X) and vertical (direction Y). The panels were subdivided in frames with 0.50 m × 0.50 m. Panels P11 and P12 were composed by a meshes 3 × 3, while the other panels were divided into meshes 2 × 3.

In the horizontal direction (direction x) the lower left corner of each frame was considered as point 0, while for the vertical direction, the sidewalk level was considered point 0, so that all panels were started at 0.5 m from it.

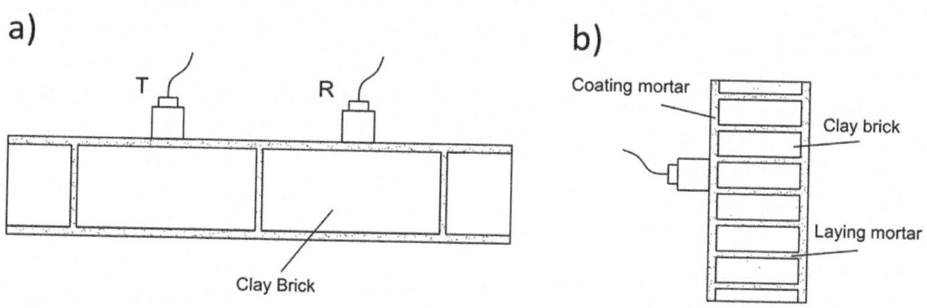

Fig. 3 Schematic view of the experimental setup: **a** top view and **b** lateral view of the ultrasonics transductors arrangement

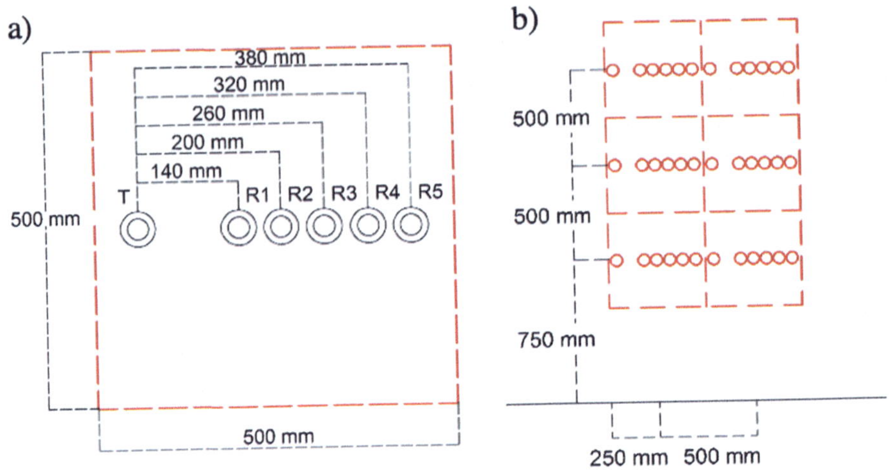

Fig. 4 Schematic view of the measurement setup: **a** distance between transmitter (T) and receivers (R) transductors; **b** scheme of frames distribution in a meshes 2 × 3

In each data acquisition frame 5 UPV measurements were done, distributed in the center of each quadrant, at heights 0.75, 1.25 and 1.75 m, with the ultrasonic receivers spaced 60 mm from each other in the horizontal direction and the transducer at a fixed distant point 140 mm from the first receiver, as shown in Fig. 4.

The distance between the transducer (T) and the first receiver (R1) was 140 mm and, sequentially, 60 mm were added to each signal reception point (R2, R3, R4 and R5), as shown in Fig. 4. The total distance between transducer T and receiver R5 was of 380 mm. After five ultrasonic velocities recorded, in each frame, the collected data were subjected to statistical treatment and the maximum and minimum values of each of the walls were identified, as well as the average speeds and their respective standard deviation, according to the work performed by [14].

4 Results

At 0.75 m of high, the mean of the UPV in the 16 walls was 1052.16 m/s, while at 1.25 and 1.75 m the values of mean of the UPV were 786.72 m/s and 726.31 m/s, respectively. Taking as reference the values of UPV at 0.75 m, the values of UPV at 1.25 and 1.75 m presented a reduction of 25.23 and 30.97%. Similar behavior of the UPV also were verified in [9, 15].

In a general way, when the clay bricks masonry walls have the coating layer well adhered, without rigid elements of openings presence, the ultrasonic wave have higher values in the region of the base of the panel, gradually decreasing as the panel advances in height. This behavior occurs face of the weight load concentration in the base of the panels, influencing the mortars porosity, and more high values of the ultrasonic velocity propagation, as also observed in the work reported by [16]. Thus, concerning the UPV distribution along a wall, the waves tend to present a velocity mean decreasing as the measurement points rise in height. The typical behavior of the UPV for a panel with coating layer well adhered can be seen by Fig. 5, where is possible to identify the UPV variation over the height.

Considering that the mortar coating is well adhered in the panel, the UPV tend to vary between 1200 and 3000 m/s, according with material density. The higher the density, the greater the ultrasonic velocity. In masonries with lime base mortar, the UPV values tend to be near to 1200 m/s, while to masonries with clay bricks and cement base mortar, the UPV present most high values, near to 3000 m/s. In lime base mortars, generally the ultrasonic waves can vary between 600 and 1000 m/s due to the high porosity of the material. Therefore, as the ultrasonic wave propagates along a well-adhered masonry panel, the velocity will present high values higher, whereas at points where there is a detachment of the coating layer, the ultrasonic wave tends to have the velocity propagation close to the UPV of the mortars. It should also be noted that the irregularity of the coating layers and the presence of cracks attenuates

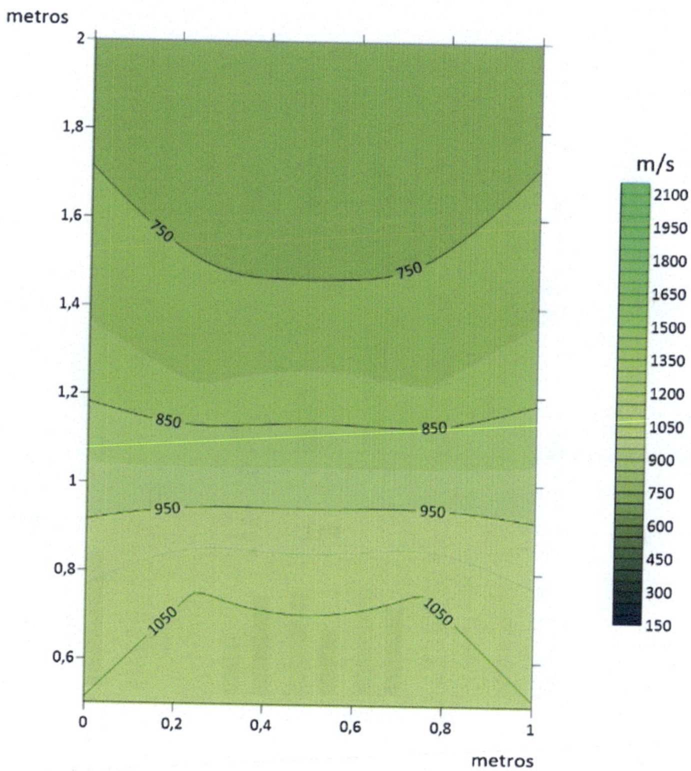

Fig. 5 Standard distribution of UPV along a masonry panel

the UPV, making possible to observe values lower than 600 m/s. This understanding concerning the behavior of VPU in masonry is observed in [16–18].

The ultrasonic velocities collected in the external walls from northeast façade, in the panel P3, as showed by Fig. 6a, was observed that the UPV average in the X directions vary from 458.35 to 677.19 m/s. To the heights of 0.75 and 1.25 m, the ultrasonic measures present similar behavior. To the height of 1.75 m, the variation in the X direction is inverse, changing from 1400.5 to 600.48 m/s. For the variation due to height, the UPV average was 1000.49 m/s in the height of 0.75 m, while for 1.25 m and 1.75 m of height the averages were 820.39 m/s and 544.32 m/s, respectively. These values represent a decrease in the ultrasonic velocities of 18 and 45.59% for 1.25 and 1.75 m of height.

For panel P4, Fig. 6b, no significant UPV variations in the X direction were detected. However, a different behavior from the other panels was detected along of the height. The ultrasonic velocity average was 635.08 m/s until the 0.75 m oh height, while to 1.25 and 1.75 m of height, the averages were 958.27 m/s and 531.23 m/s, respectively. These values represent an increase in the ultrasonic velocity

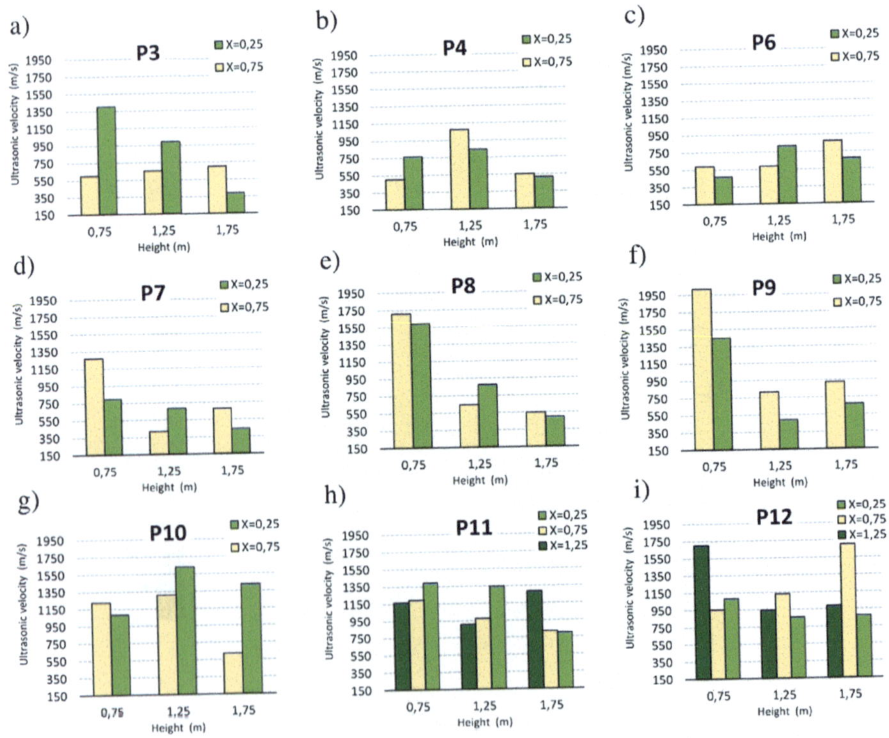

Fig. 6 Influence of the height on UPV measured in the Northeast facade of the Nossa Senhora da Expectacão Church

from 50.89% to 1.25 m, followed by a decrease from 16.35% to 1.75 m in height. Both changes had as reference the average speed at 0.75 m of height.

In the P6 panel (Fig. 6c), the range of ultrasonic velocity observed varies between 526.65 m/s to 0.75 m in height, to 702.26 m/s and 771.42 m/s to 1.25 m and 1.75 m respectively. These values variation represent an increase in the ultrasonic velocities of 33.35 and 46.48% for 1.25 and 1.75 m in height in comparison with point measured at 0.75 of height. The behavior UPV identified in this panel is opposite to the expected, which would be the decrease the velocities while the height rises. A visual inspection carried out in this panel region showed cracks presence in the top region of the panel.

Panel P7 (Fig. 6d) showed a more significant variation of UPV in the X direction, especially at 0.75 m of height where the VPU varied from 796.09 to 1268.60 m/s. In reference to the height, the ultrasonic velocity was 1032.35 m/s at 0.75 m, while for 1.25 and 1.75 m the velocities were 550.28 m/s and 558.45 m/s, respectively. These values represent a decrease in the ultrasonic velocities of 46.70 and 45.9% for 1.25 and 1.75 m of the height.

Panel P8 (Fig. 6e) presented an UPV distribution very uniform. The UPV was of 1648.05 m/s at 0.75 m of height, while for 1.25 and 1.75 m the velocities were 756.25 and 518.71 m/s. These values represent a decrease in the ultrasonic velocities of 54.11 and 68.53% to 1.25 and 1.75 m in height in comparison with point measured at 0.75 m.

The range of ultrasonic velocities observed in P9 (Fig. 6f) varied between 2017.60 and 490.19 m/s, a variation greater than in the other panels. In this panel, was possible to observe a significant variation in the X direction, mainly at the 0.75 m of height, where the UPV varied from 1447.80 to 2017.60 m/s. The UPV average measured at 0.75, 1.25 and 1.75 m of height were 1732.70, 651.44 and 795.30 m/s, corresponding to a rate of decrease in the ultrasonic velocity of 62.40 and 54.10%, considering the reference point at 0.75 m.

The panel P10 (Fig. 6g) presented a significant variation at X direction, especially at 1.75 m, where the UPV varied from 1424.80 to 621.41 m/s. The UPV measured was of 1153.55 m/s at a height of 0.75 m, while for 1.25 and 1.75 m of the height the UPV presented values of 1467.00 and 1023.11 m/s. These values represent an increase in the ultrasonic velocity of 27.17% followed by a decrease from 11.31% to 1.25 m and 1.75 m in height, considering the point measured at 0.75 m of height as reference. This panel presented results outside of the expected values once it is noticed that panel P10 presented velocities peaks at the heights of 1.25 and 1.75 m.

In order to analyze the influence of a wall connected perpendicularly in an analyzed panel, Panels P11 and P12 were chosen. This configuration is present in these two panels and can be seen by Fig. 2.

Panel P11 (Fig. 6h) has a quite slightly different configuration from the others. Panel P11 presents a larger mesh, given the availability of space to perform the test. P11 showed some variations in the ultrasonic velocity in the X direction without a known pattern distribution regarding heights. At the height of 1.25 m, the velocities varied between 1338.30, 972.57 to 906.57 m/s. At the height of 1.75 m, ultrasonic velocities varied from 794.74, to 816.02 m/s and sequentially to 1282.10 m/s. Now, the averages of the UPV variations along the height showed an expected behavior. The average of the ultrasonic velocity measured was 1248.13 m/s at a height of 0.75 m, while for 1.25 m and 1.75 m the UPV collected were 1072.48 m/s and 964.29 m/s, respectively. These values represent a decrease in the ultrasonic velocities of 14.07 and 22.74% for 1.25 and 1.75 m in height.

Panel P12 (Fig. 6i) has a similar configuration than the observed in P11, presenting similar results. Variation at the X directions also were noted as observed in P11. For height 0.75 m, the velocities in the positive direction of X varied from 1077.80, 952.11 to 1700.50 m/s. At the height of 1.75 m, UPV varied from 871.88, 1695.3 to 1700.50 m/s. For the variation due to height, the UPV measured were of 1243.47 m/s to 0.75 m, while for 1.25 m and 1.75 m the collected velocities were 976.67 m/s and 1185.28 m/s, respectively. These values represent a decrease in ultrasonic velocities of 21.46 and 4.68% to 1.25 and 1.75 m in height.

For both panels P11 and P12, it was possible to notice a significant variation in the X direction. The observed behavior is due to a wall presence connected perpendicularly in the analyzed panel. Thus, some quadrants overlapped closer to the attachment region resulting in UPV higher than those adjacent to the sides. As results, a noticeable changing in the UPV behavior at the X direction.

Analyzing the external walls of the Southwest facade of the Church, panel P13 (Fig. 8a) presented a very unexpected behavior. In the X direction it showed a greater variation at the 1.75 m of height, varying from 841.47 to 1229.70 m/s in the positive X direction. The UPV measured were of 792.87 m/s for a height of 0.75 m to 1.25 m and 1.75 m, were 885.03 m/s and 1035.59 m/s respectively. The opposite behavior of what was expected. In this panel the ultrasonic velocities had an increasing behavior as the heights rise. These velocities values correspond to a percentage increase of 11.62 and 30.61%, taking 0.75 m as reference. As this wall does not have any special configuration or visual cracks, and these results may indicate the presence of voids at the bottom region of the panel.

Panel P14 (Fig. 8b) showed variations in the X direction. In the 0.75 to 1.25 m, the velocities increased in an order of 300 m/s. While following in the positive direction of X, in the 1.75 m the UPV varies from 1073.30 to 735.12 m/s. The UPV average was 1316.8 m/s at a height of 0.75 m, while for 1.25 m and 1.75 m the averages were 1067.26 m/s and 904.21 m/s, respectively. These values represent a decrease in ultrasonic velocities from 1.95 and 31.33% to 1.25 and 1.75 m in height.

In the panel P16 (Fig. 8c), the UPV measured were 1123.75 m/s at a height of 0.75 m, while for 1.25 and 1.75 m the ultrasonic velocities collected were 782.03 and 869.23 m/s. These values represent a decrease in ultrasonic speeds from 30.41 and 22.65% to 1.25 and 1.75 m in height, with the standard reference. With these results, it was noticed an area with lower speeds at a height of 1.25 m, which may indicate discontinuities in the panel at this point.

Panel P17 (Fig. 8d) showed results that are quite different from the others analyzed panels. First, their UPV showed low values in comparison to the other panels, with a maximum speed of 863.33 m/s. For the variation in the X direction, it is found that at the 0.75 m elevation the ultrasonic velocity ranges from 806.21 to 458.94 m/s, for the 1.25 m of height UPV varied from 863.33 to 425.65 m/s, at the 1.75 m height there are no great variations. For the variation due to height, the ultrasonic velocity was

632.58 m/s up to 0.75 m, while for 1.25 m and 1.75 m the averages were 644.49 m/s and 560.52 m/s, respectively. These values represent an increase in the UPV of 1.88% followed by a decrease of 11.39% for heights 1.25 and 1.75 m.

Panel P18 (Fig. 8e) presented greater variation in the X direction at height 0.75 m, from 1425.3 to 993.68 m/s. For the UPV in relation to height, the velocities were of 1209.49 m/s was obtained for 0.75 m, for 1.25 m and 1.75 m the velocities were 421.91 m/s and 416.86 m/s, respectively. This variation corresponds to a percentage decrease of 65.12 and 65.53% at heights 1.25 and 1.75 m, taking the point 0.75 m as reference. Although the panel presented the expected behavior, the variations referring to heights presented great amplitudes, with the highest dimensions, low velocities of ultrasonic pulse propagation compared to the other panels, indicating an influence of the loading at the panel behavior.

Panel P19 (Fig. 8f), also showed greater variation in the X direction at the 0.75 m elevation, where the velocities varied from 635.79 to 1093.4 m/s. The UPV were 864.6 m/s for the 0.75 m of height, 198.79 m/s for 1.25 m and 371.19 m/s for the 1.75 m of height. This variation corresponds to a percentage decrease of 77.01 and 57.07% at heights 1.25 and 1.75 m.

Considering the comparison between all UPV values from all analyzed panels, for different level of height, can be seen that the low values of ultrasonic velocities indicate the presence of a damaged in the panel, which can range from cracks in the coating mortar to non-adherence. The presence of cracks in the coating mortar, as showed by Fig. 7, was a common factor in all panels that presented the above commented UPV behavior.

The P20 panel (Fig. 8g) showed visible cracks in the mortar, as well as it can also be verified by a brief percussion test performed on the panel. Thus, the panel results interpretation was made difficult, the voids formed by the cracks make the panel inhomogeneous, and the presence of these voids reduces the propagation speeds in the masonry. In this way, the results tend to present low speeds and without great

Fig. 7 Panel P19 with cracks generalized along all panel

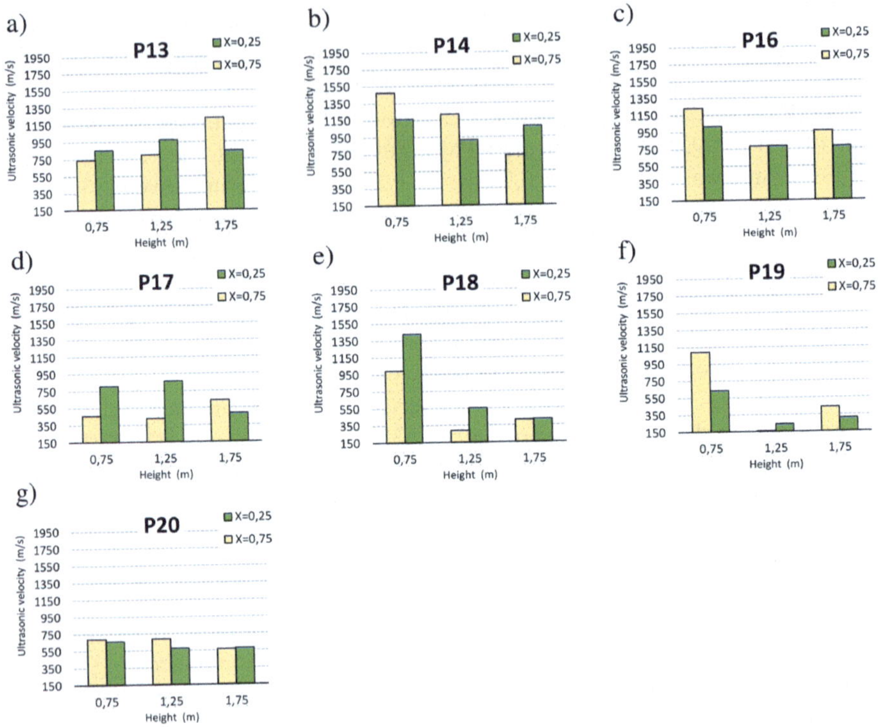

Fig. 8 Influence of the height in the UPV of the Southeast facade of the Nossa Senhora da Expectacão Church

variations with the position. Thus, there were no major variations in the X direction, the UPV remained practically constant. An UPV of 674.07 m/s was obtained for 0.75 m of height, for the 1.25 and 1.75 m elevations the collected velocities were 632.98 m/s and 571.30 m/s respectively. This variation corresponds to a percentage decrease of 6.10 and 15.25% at heights 1.25 and 1.75 m, considering point 0.75 as reference.

Considering the values of UPV measured in all assessed panels, Table 1 shows the maximum and minimum values for everyone panel. Beyond that, the UPV averages and the standard deviation were obtained.

However, only based on the UPV variation present by Table 1 it is not possible to detect the effects of the coating layer detachment or cracks. The high variation between the UPV from the same panel, indicates a higher level of heterogeneity. For identify the influence of cracks or coating detachment, the pattern of UPV distribution need to be considered, as showed by Fig. 9.

The general analysis of the maximum, minimum and average values makes possible to identify that the panels P6, P17, P19 and P20 showed velocities below of the UPV average of all panel, indicating possible damage presence. It can also be identified that the panel P20 presented a very low standard deviation, which indicates

Table 1 General results of UPV of the Nossa Senhora da Expectacão Church

Ultrasonic velocity (m/s)	Maximum	Minimum	Average	Standard deviation
P3	1400.50	388.29	788.40	356.43
P4	1073.80	503.71	708.19	228.18
P6	871.36	466.50	666.77	153.78
P7	1268.60	417.47	713.69	309.46
P8	1706.20	493.40	974.33	539.00
P9	2017.60	490.19	1059.81	570.16
P10	1629.20	621.41	1214.55	344.75
P11	1387.40	794.74	1094.97	227.09
P12	1700.50	854.29	1135.14	330.92
P13	1229.70	737.80	904.49	177.04
P14	1471.30	735.12	1096.09	253.97
P16	1230.20	776.96	925.00	182.10
P17	863.33	425.65	612.53	187.51
P18	1425.30	290.65	682.75	438.19
P19	1093.40	159.49	478.19	344.69
P20	688.54	565.09	626.11	58.92
Average	1316.06	545.05	855.06	293.89

a homogeneity in the velocity values. The low variation between the measured points may point to presence of damage. In panel P20 high levels of crack in the coating mortar were identified.

The map of UPV distribution for the P3 panel (Fig. 10) showed a high variation in the X direction found at the base of the panel. The most accentuated variation occurs at height of 0.75 m. Such variation shows the low velocity zone of the right bottom part of the panel, approximately in X = 0.80 m.

Following, the mapping of the ultrasonic velocities of the P4 panel (Fig. 11) shows values below 1100 m/s. In addition, the panel base has lower values of velocities than other areas, where the UPV average was 635.08 m/s at a height of 0.75 m, quite distant than result from the panel average of 1052.16 m/s. These results show areas of low homogeneity in the panel masonry or damage in the coating mortar in this region. The UPV map also shows low-speed areas at the top of the panel at a height of 1.75 m.

The UPV mapping of the P6 panel (Fig. 12) presented a behavior very similar to P4. A low average of the ultrasonic speeds was obtained. A highlight can be attributed to ultrasonic average at the 0.75 m of height, with average of 526.65 m/s. These results indicate that this panel presence of voids or cracks.

From the analysis of the UPV distributions in panel P7, shown by Fig. 13, the variation of ultrasonic velocities is observed in both directions. The maximum velocity found was 1268.60 m/s located in the bottom region of the panel.

For panel P8 and P9 (Figs. 14 and 15, respectively), a very similar mapping was

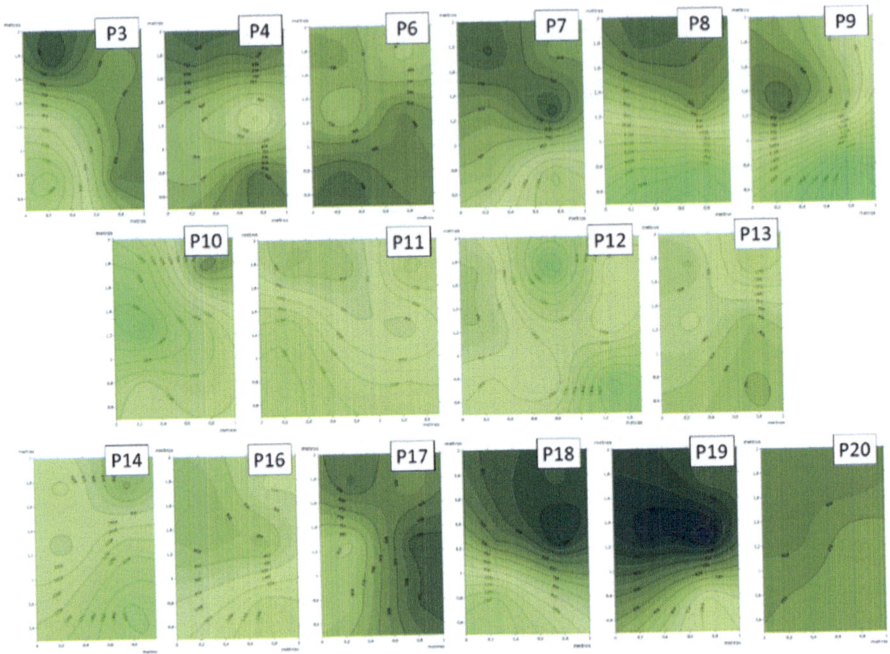

Fig. 9 Ultrasonic velocity mapping of the masonry panels of the Nossa Senhora da Expectacão Church

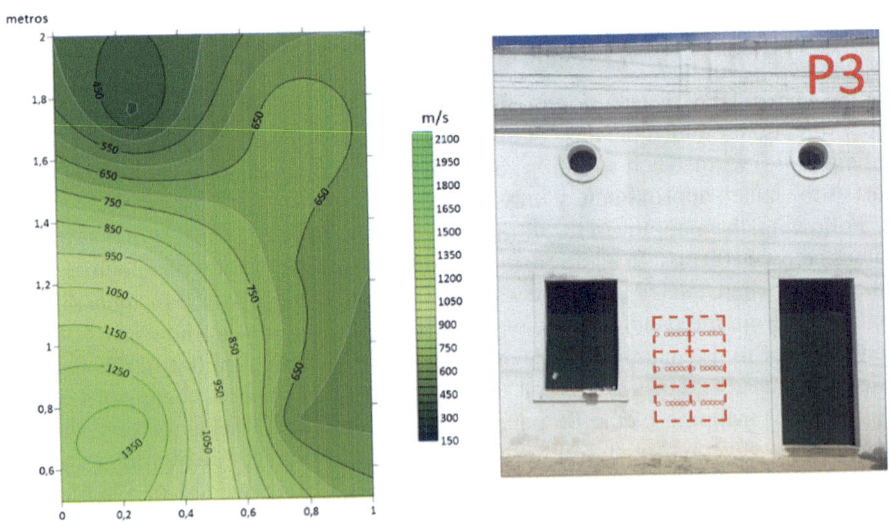

Fig. 10 UPV mapping of the panel P3

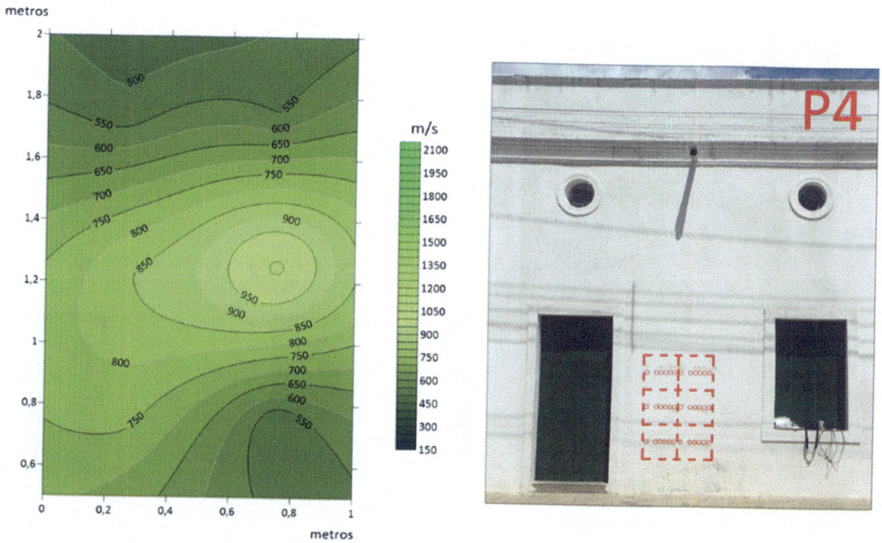

Fig. 11 UPV mapping of the panel P4

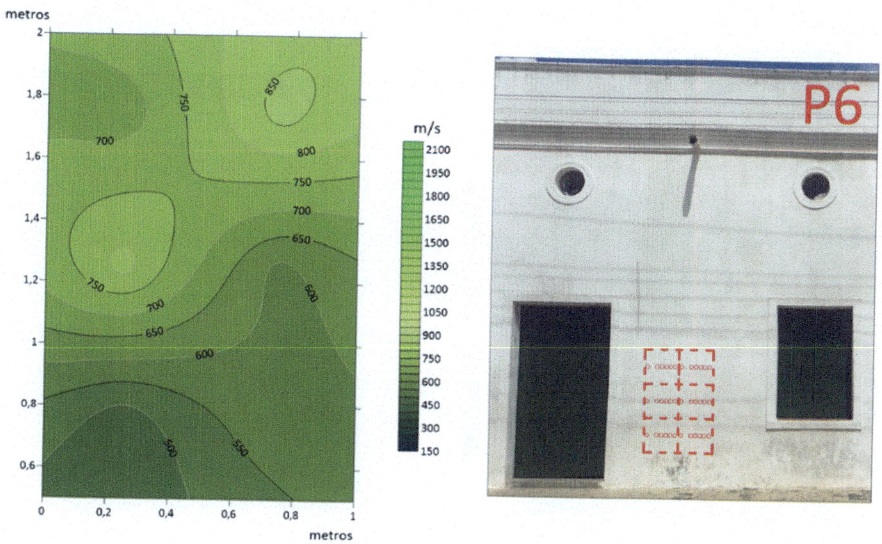

Fig. 12 UPV mapping of the panel P6

obtained. Both panels presented high values of UPV at the base of the panels, namely 1648.05 and 1732.70 m/s. These two panels showed the highest values for standard deviation around 550 m/s. The obtained results can be associated with the gradual and uniform distribution of the variation of the UPV. However, for panel P9, at 1.25 m

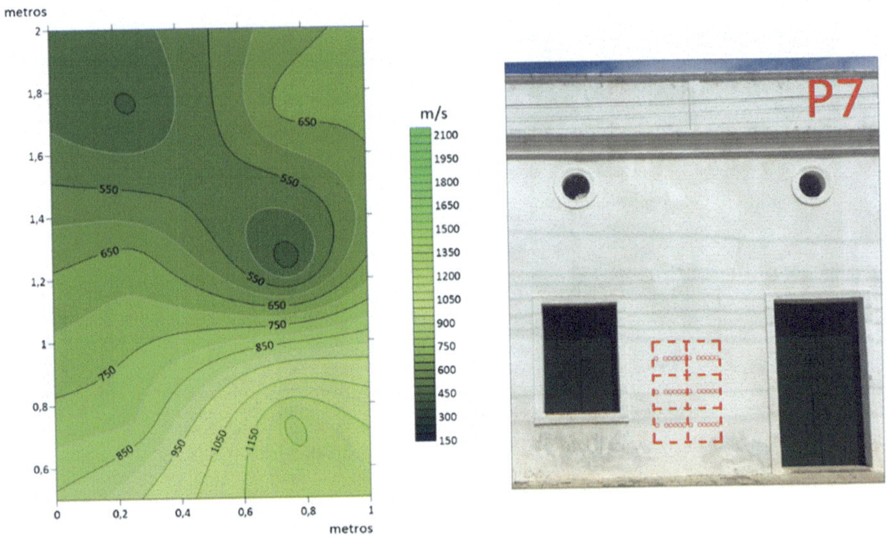

Fig. 13 UPV mapping of the panel P7

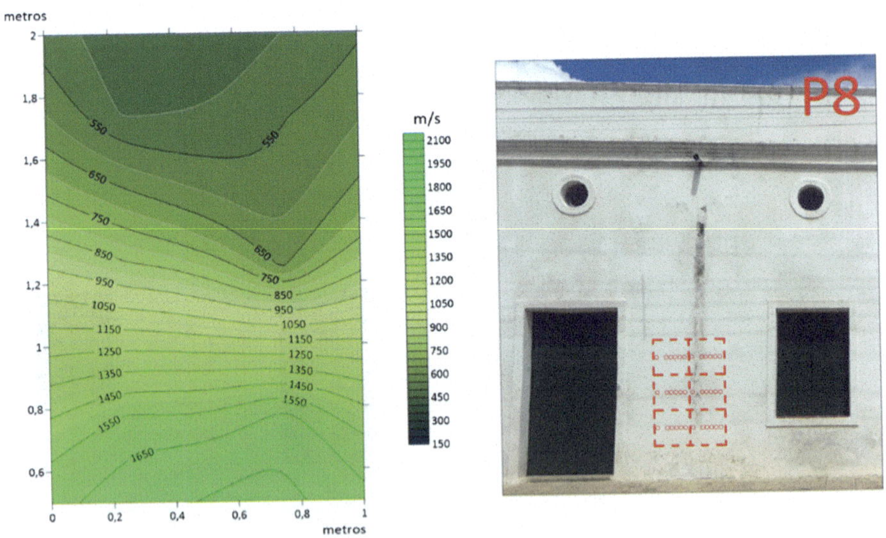

Fig. 14 UPV mapping of the panel P8

of height was found a velocity of 550.00 m/s, showing the possibility of damage located in this region.

When analyzed the ultrasonic mapping performed for the P10 panel (Fig. 16), a different behavior can be seen from the others, with high values of UPV distributed

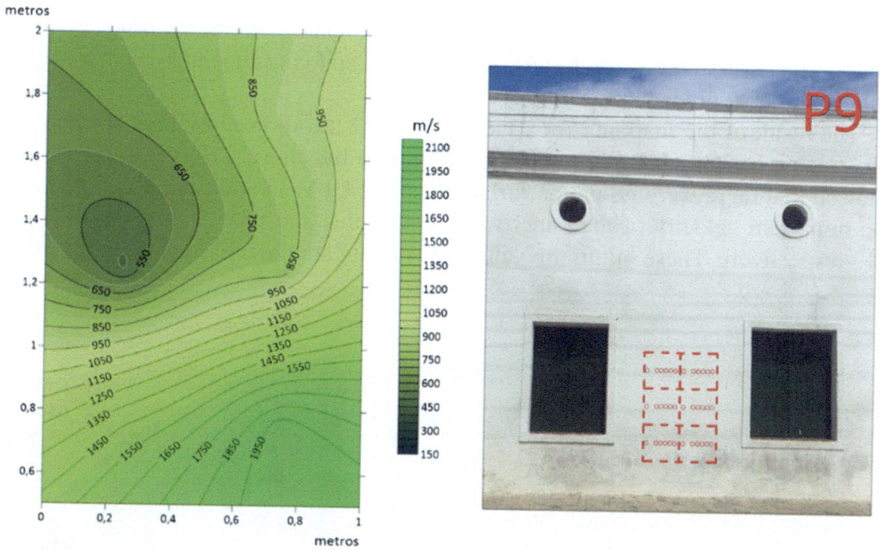

Fig. 15 UPV mapping of the panel P9

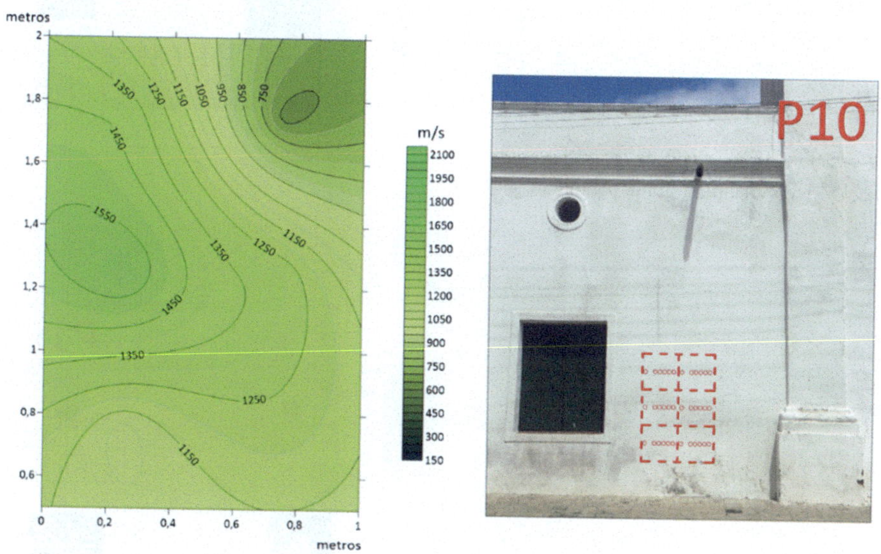

Fig. 16 UPV mapping of the panel P10

throughout the panel with a maximum of 1629.20 m/s located at a height of 1.25 m. This velocity distribution can be explained by two peculiar factors to this panel: the panel is located next to the columns of the church tower and the back of the panel is attached to the wall of the stairs that gives access to the leather and the church tower (Fig. 3). This configuration contributes to a change in the rearrangement of the compression and tension forces of this panel, and consequently causing higher values of ultrasonic velocity.

The result of the mapping for panels 11 and 12 are showed by Figs. 17 and 18. This region has an intercommunicated wall inside the church. However, the results showed that there were no major variations in the UPV. In addition, it is noted that the minimum velocity of the panel were 794.74 m/s and 854.29 m/s, for P11 and P12 respectively. These minimum values of velocity are the highest among the other

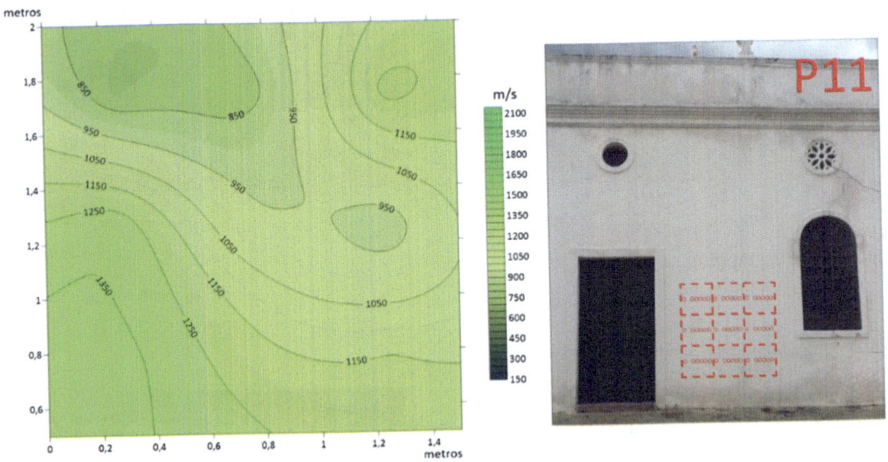

Fig. 17 UPV mapping of the panel P11

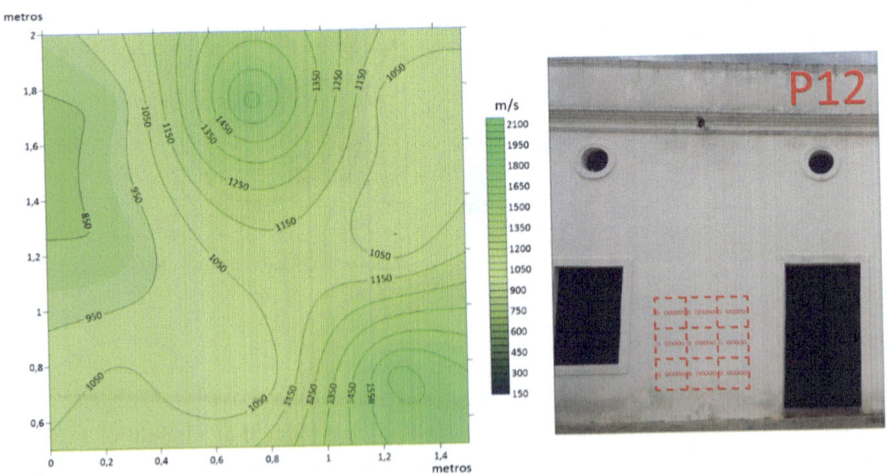

Fig. 18 UPV mapping of the panel P12

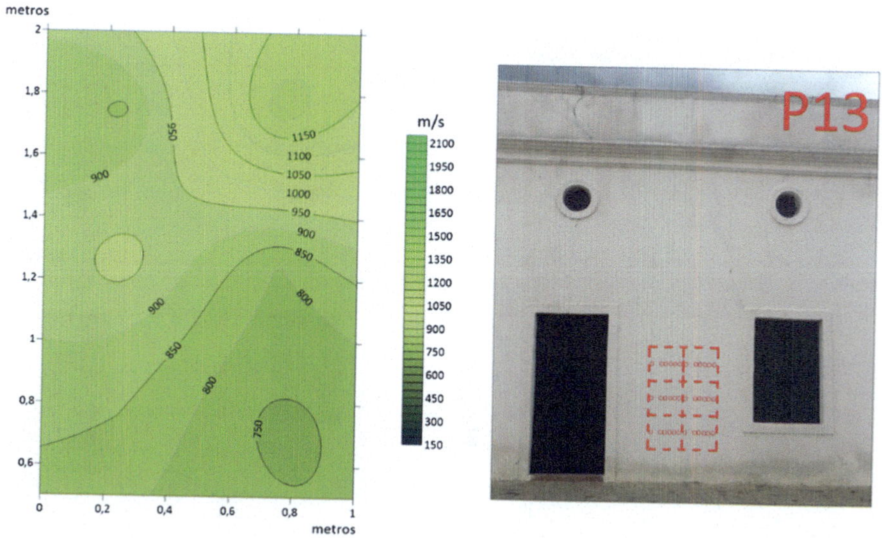

Fig. 19 UPV mapping of the panel P13

panels. In this way the influence caused by the presence of the wall connected to the panel is verified, as previously mentioned.

For the mapping of panel P13 (Fig. 19) it is possible to verify an area with lower velocities at the bottom of the panel, as well as the presence of areas with higher speeds at the top.

The mapping of panels P14 and P16 are presented by Figs. 20 and 21. The main difference between both is in the maximum velocity values found, which were 1471.30 m/s and 1230.20 m/s respectively. In addition, small variations along the X direction also can be noted for both panels.

The mapping of Panel P17 (Fig. 22) presents a better model of the UPV behavior along the panel. Panel P17 shows a wide variation in ultrasonic velocities in the X direction and values well below the averages. A hypothesis already presented is that P17 presents a structural failure, possibly a crack, causing a disconnection between two regions and, consequently, different velocities in the same panel. This behavior can also be linked to the application of different materials, or even from different ages, resulting in differences in the velocities on the same panel separated by a vertical line. The panel also has an average velocity of 612.53 m/s, below from the average velocities of the other panels, over around 855.06 m/s.

For the mapping of the P18 panel (Fig. 23), it is possible to verify a local behavior with low velocities values, close to 350.00 m/s. This region is located in the right side of the panel around 1.75 m. It is also possible to notice a great amplitude in their velocities. The minimum velocity found in this panel was 291 m/s, well below the UPV minimum of 545.05 m/s. In association with UPV values of P18, a high standard deviation was found, namely 438 m/s. Comparing this value with an average

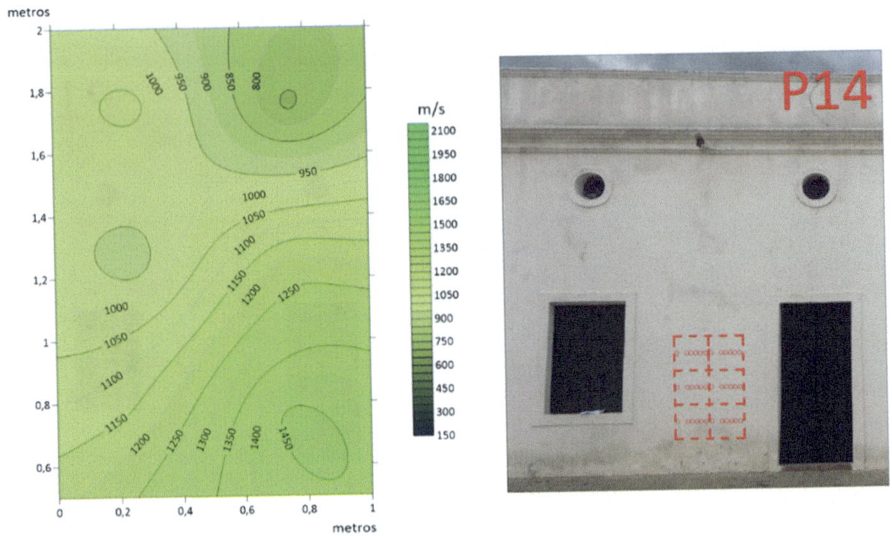

Fig. 20 UPV mapping of the panel P14

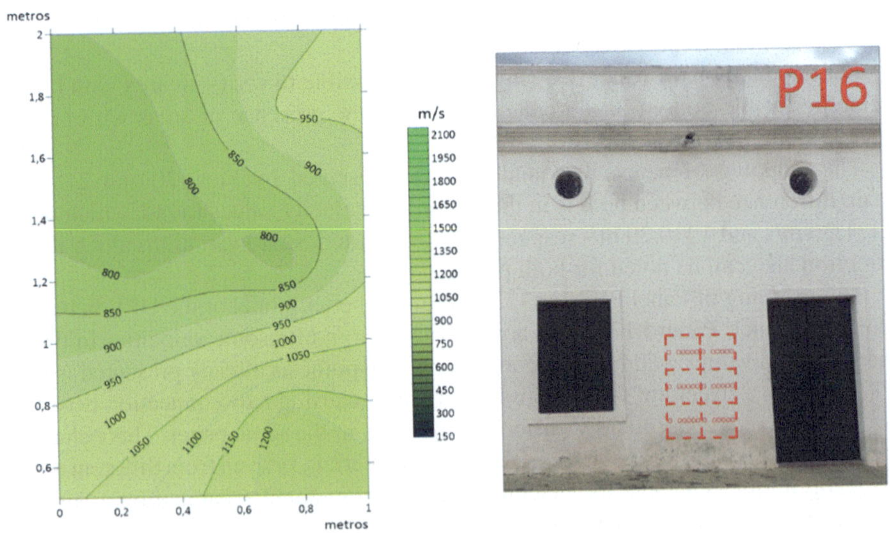

Fig. 21 UPV mapping of the panel P16

of 293.89 m/s, these results can be highlighted as an indicator of cracks in the mortar located in a region of the panel.

Panels P19 and P20 deserve greater attention, since during the data acquisition was verified that they had a superficial cracking. This cracking was identified visually and by a brief percussion test.

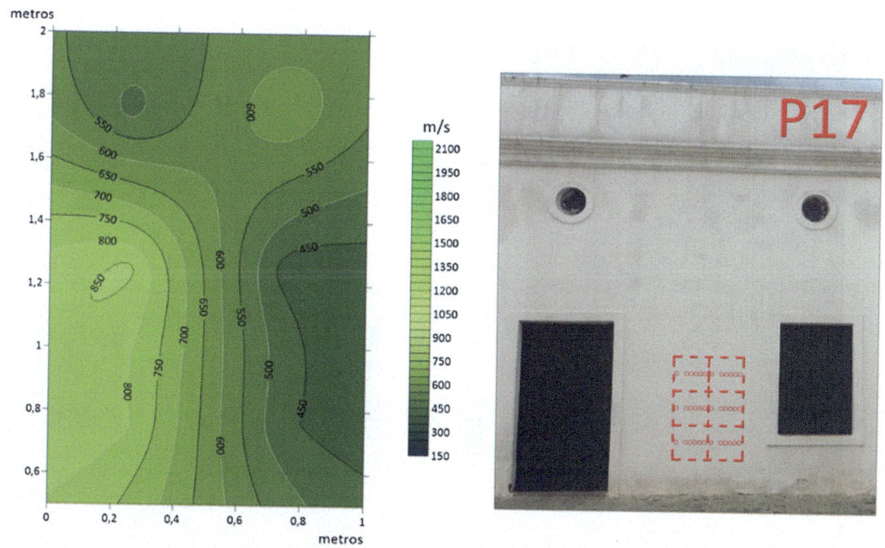

Fig. 22 UPV mapping of the panel P17

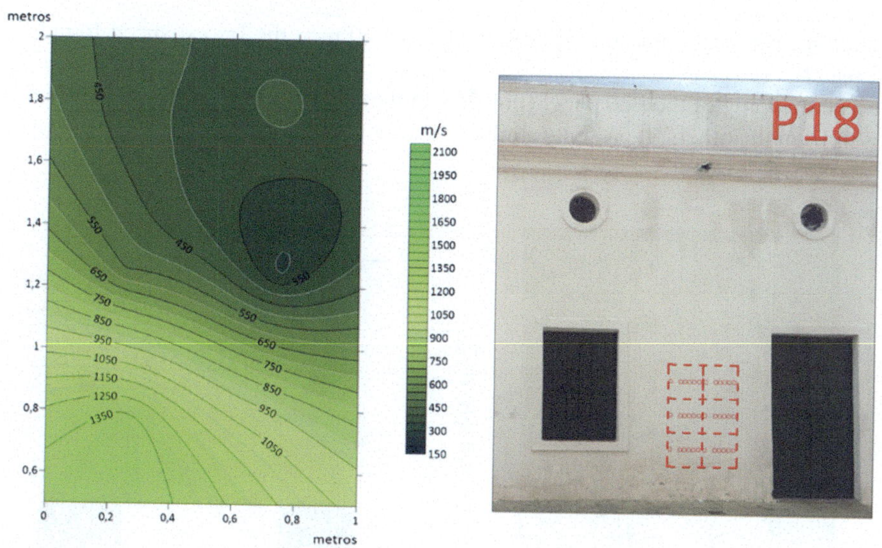

Fig. 23 UPV mapping of the panel P18

For panel P19 (Fig. 24) there was only a local crack identified around the 1.25 m of height. This result can be seen in the mapping of P19 panel, where an area with lower velocity values at height of 1.25 m is noticeable. This UPV local behavior is consistent with the area where cracks in the coating mortar were verified in situ. The

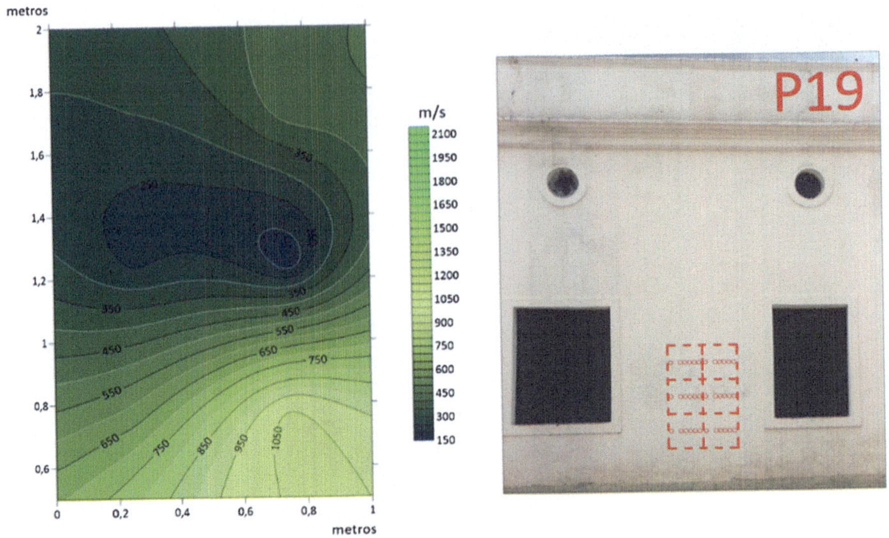

Fig. 24 UPV mapping of the panel P19

minimum velocity found on this panel was 159.49 m/s, the lowest speed found on all panels.

The mapping of the P20 panel (Fig. 25) on the other hand, shows a very different result from the others. It presents almost constant ultrasonic velocities over its entire

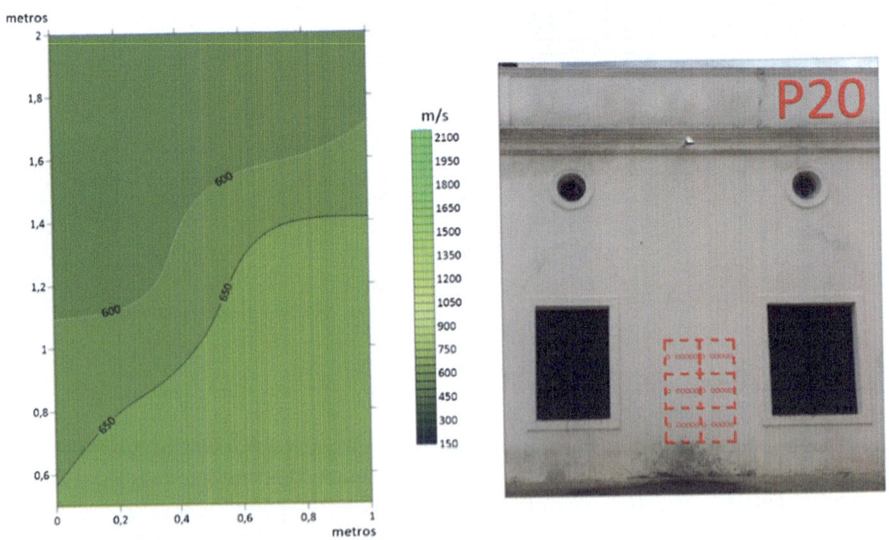

Fig. 25 UPV mapping of the panel P20

length, with an UPV average of 626.11 m/s. This result is a reflection of the cracking found in this panel. The panel presents cracks extending along its area. An explanation for this constant behavior is that the mortar is not well adhered to the masonry. So, in this case the waves propagate only in the layer of the mortar coating, resulting in this constant behavior throughout the panel, as showed by Fig. 26. When analyzing Table 1, it is possible to see that the P20 panel, presented the lowest maximum velocity among the other panels, with 689 m/s and also presenting a standard deviation well below to the others, with only 58.9 m/s, and an UPV average of 293.89 m/s.

In addition, in the masonry areas identified as non-adhered coating by ultrasonic tests a cave sounds were noted, as well pieces of the coating mortar detached easily when submitted to a hammer impact.

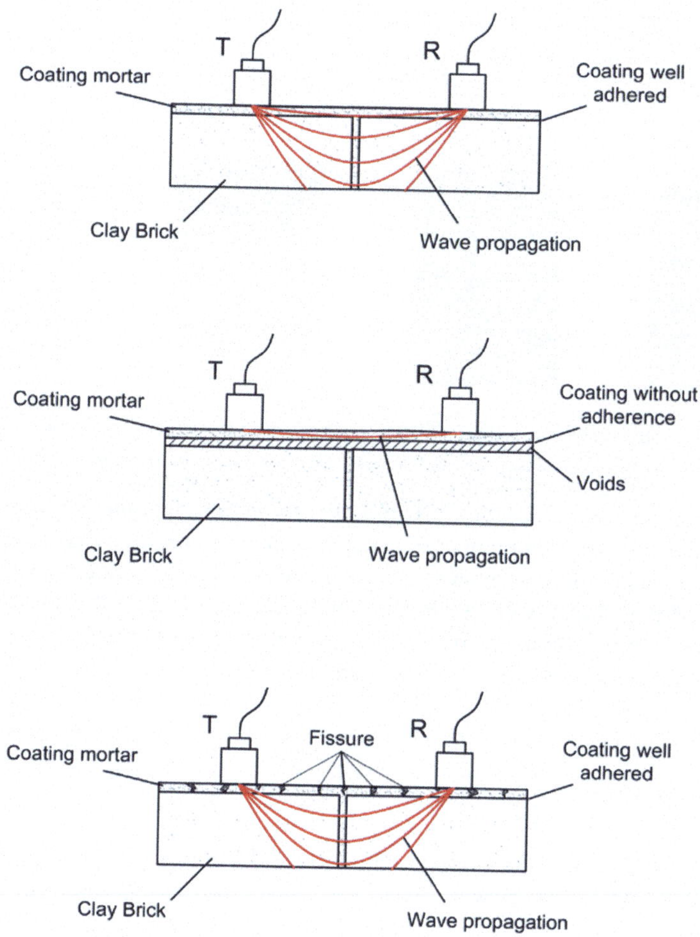

Fig. 26 Schemes of ultrasonic waves propagation according coating mortar adherence quality

Fig. 27 UPV mapping of the Northeast facade of the Nossa Senhora da Expectacão Church

Analyzing the lateral façades, it is possible to observe that the Northeast façade (Fig. 27) presents a greater number of panels within the expected behavior, even with variations in panels close to other structural elements, such as panels 4 and 6 that are close to a perpendicular wall. On the Southwest façade (Fig. 28), the results point to a higher incidence of damage, mainly along panels P17, P19 and P20. Furthermore, there are greater variations in the UPV along this facade.

5 Conclusions

In this work the methodology proposed by Mesquita et al. (2018) and published as ABNT 16805 (ABNT, 2020) was used to characterize the masonry walls of the Nossa Senhora da Expectação Church. In general, 16 walls from Northeast and Southeast facades were analyzed.

Fig. 28 UPV mapping of the Southeast facade of the Nossa Senhora da Expectacão Church

A comparison between the walls allowed to note that the walls with presence of cracks influences significantly in the UPV. The variation range of UPV is largest in the damaged walls.

Considering that each material presents a characteristic velocity of ultrasonic propagation according to density, the existence of damage, voids, or other heterogeneities can influence the pattern of UPV. Thus, the employment of UPV to characterize masonries properties can be a useful tool to assessment of historic buildings.

Beyond of the profile distribution of ultrasonic waves propagation in masonries can be used to identify anomalies in a wall, these waves profiles distribution also can be used to identify similarities along walls from the same masonry structure and indicates if all walls used the same constructive methodology. In fact, this is a good tool to historical analysis, allowing to see if an ancient building was completely built with the same constructive method or if some of the walls presents differences between them.

In addition, the results of this work indicate typical behavior of the UPV in masonries walls built with vernacular clay bricks joined by a lime mortar and lime mortar coating with a thickness not exceeding 1.5 cm, specifically:

- When submitted to weight load, without confinement effect or openings, ultrasonic waves present velocity of propagation higher in the base of the wall, decreasing according to the height advance;

- To the case of non-generalized cracks presences in the coating layer, the regions with cracks presence presents UPV near of the typical values of the ultrasonic velocities of lime mortars (over around 600 m/s or lower than that for deep cracks), while in the other areas of the walls the typic pattern of UPV distribution will be noted;
- To the case of generalized cracks conditioning the adherence between the coating layer and the wall, the ultrasonic waves presents a pattern of propagation almost constant, with low range of variation, essentially between 600 m/s and near to 1000 m/s—this values can be attenuated by presence of voids and cracks and can vary according to mortar density;
- The existence of heterogeneities (structural faults or changes in the material) can be detected by abrupt change in the ultrasonic velocities, from the pattern of distribution to a lower value (down of 600 m/s);
- The presence of rigid structural elements as columns or walls connected to the assessed wall tends to influence the increasing of the UPV values, making the velocities vary around 1500 m/s and may present higher values.

Finally, in this paper a new point of view on UPV interpretation was introduced, namely based on analysis the pattern of waves distribution. Along of the past decades, the literature reported ultrasonic analysis based on consideration of ultrasonic velocities only, without consider the arrangement between the individual ultrasonic values or the influences of heterogeneities to the ultrasonic waves behavior. The observation of UPV pattern makes possible the better characterization of heterogeneous materials as masonries, and the establishment of a waves distribution profile, open a new way to studies on masonries characterization, especially the historic ones.

References

1. E. Mesquita, P. Antunes, F. Coelho, P. André, A. Arêde, H. Varum, Global overview on advances in structural health monitoring platforms. J. Civ. Struct. Heal. Monit. **6**(3) (2016). https://doi.org/10.1007/s13349-016-0184-5
2. N. Kurata, B.F. Spencer, M. Ruiz-Sandoval, Risk monitoring of buildings with wireless sensor networks. Struct. Control Heal. Monit. **12**(3–4), 315–327 (2005). https://doi.org/10.1002/stc.73
3. E. Mesquita et al., Heterogeneity detection of Portuguese-Brazilian masonries through ultrasonic velocities measurements. J. Civ. Struct. Heal. Monit. (2018). https://doi.org/10.1007/s13349-018-0312-5
4. L. Binda, A. Saisi, State of the Art of Research on Historic Structures in Italy (2001), 1996
5. E. Mesquita, A. Cavalcante, L. Mota, E. Araújo, A. Diógenes, Caracterização dos danos em duas construções históricas de alvenaria vernacular da cidade de Sobral (Damage characterization of Two heritage buildings of vernacular masonry from Sobral City) (2016), pp. 1–14
6. ICOMOS, "Icomos Charter- Principles for the analysis, conservation and structural restoration of architectural heritage," in International Council on Monuments and Sites, 2003
7. L. Binda, A. Saisi, L. Zanzi, Sonic tomography and flat-jack tests as complementary investigation procedures for the stone pillars of the temple of S. Nicolò l'Arena (Italy). NDT & E Int. **36**, 215–227 (2003). https://doi.org/10.1016/s0963-8695(02)00066-x
8. L.F.B. Miranda, Ensaios acústicos e de macacos planos em alvenarias resistentes (Universidade do Porto, 2011)

9. L. Binda, A. Saisi, C. Tiraboschi, Application of sonic tests to the diagnosis of damaged and repaired structures. NDT & E Int. **34**, 123–138 (2001)
10. L.U. Binda, A. Saisi, C. Tiraboschi, Investigation procedures for the diagnosis of historic masonries. Constr. Build. Mater. **14**, 199–233 (2000)
11. L. Miranda, J. Guedes, J. Rio, A. Costa, Stone Masonry Characterization Through Sonic Tests, Porto (2010)
12. R. Martini, Caracterização de alvenarias de granito baseada em técnicas geofísicas, mecânicas e redes neuronais (University of Porto, 2019)
13. ABNT—Associação Brasileira de Normas Técnicas, *Non-destructive Testing—Ultrasound—Panels Characterization Through Velocity of Waves Propagation* (ABNT, Brazil, 2020), p. 28
14. E. Mesquita, R. Martini, A. Alves, P. Antunes, H. Varum, Non-destructive characterization of ancient clay brick walls by indirect ultrasonic measurements. J. Build. Eng. **19**(January), 172–180 (2018). https://doi.org/10.1016/j.jobe.2018.05.011
15. E. Mesquita, R. Martini, A. Alves, P. Antunes, H. Varum, Non-destructive characterization of ancient clay brick walls by indirect ultrasonic measurements. J. Build. Eng. **19** (2018). https://doi.org/10.1016/j.jobe.2018.05.011
16. E. Manning, L.F. Ramos, F.M. Fernandes, Direct sonic and ultrasonic wave velocity in masonry under compressive stress, in *9th International Masonry Conference* (2014), pp. 1–12
17. E. Mesquita et al., Heterogeneity detection of Portuguese–Brazilian masonries through ultrasonic velocities measurements. J. Civ. Struct. Heal. Monit. **8**(5) (2018). https://doi.org/10.1007/s13349-018-0312-5
18. P. Turgut, Properties of masonry blocks produced with waste limestone sawdust and glass powder. Constr. Build. Mater. **22**(7), 1422–1427 (2008). https://doi.org/10.1016/j.conbuildmat.2007.04.008

A Novel Approach for Detection of Voids in Traditional Load-Bearing Masonries Based on Ultrasonic Data

Talisson Rodrigues, Francisco Damasceno Filho, Rosineide Paz, Esequiel Mesquita, and Rachel Martini

Abstract This work has as the main aim proposes a novel approach for the detection of voids in ancient masonries based on ultrasonic data collected by indirect measurements. For that, two samples of ancient walls were built and tested in the laboratory. The first wall sample was built with anomalies (voids) distributed along its area, while the second one was built without voids. The wall samples were divided into 12 frames. In each frame 6 time-measures of ultrasonic propagation were performed. The results showed that the ultrasonic data collected by indirect measurements can be used to characterize masonries. Also, the results show that the assembly between Mixed effect and Lowess models for processing ultrasonic data, can be a powerful tool to detect and identify the voids inside the masonries walls, even in anisotropic materials like masonries, if the distribution of the waves and its behavior is understood, it is possible to identify disturbs on the data by anomalies presence, as voids. This work also demonstrates that, for analysis of anisotropic materials, the pattern distribution of ultrasonic waves can be an alternative way to characterize masonries.

Keywords Ultrasonic data · NDT · Masonry · Voids · Historic buildings

1 Introduction

Heritage constructions stimulating the interest of the scientific community for the development of non-destructive strategies and methodologies to support the maintenance of structural safety. Especially due to the damage mechanisms, constructive issues and its own heterogeneity, to assess masonries by non-destructives tests (NDT) has been a complex task. While since 60-decade several advances have been done in

T. Rodrigues · F. D. Filho · R. Paz · E. Mesquita (✉)
Laboratory of Rehabilitation and Building Durability, Federal University of Ceará, Campus Russas, Ceará 62900-000, Brazil
e-mail: emesquita@ufc.br

R. Martini
CEFET-MG, Civil Engineering and Environment Department, Federal Center for Technological Education in Minas Gerais, Campus Curvelo, Minas Gerais 35790-000, Brazil

J. M. P. Q. Delgado (ed.), *Case Studies in Building Constructions*, Building Pathology and Rehabilitation 15, https://doi.org/10.1007/978-3-030-55893-2_5

most isotropic materials, as steel and concrete, application on anisotropic materials as masonry are missed in the literature [1].

According to ICOMOS recommendations [2], ancient structures must be analyzed with less invasive investigation techniques and the rehabilitation measures must be carried out with the maximum efficiency and careful way, in order of do not introduces new damages to the structure. Among the usual tools for evaluating the structural characteristics, due to repeatability without introduce damages as well the necessity of sample extraction, NDT appears as a promising tool to analyze the properties of the historic buildings. With satisfactory results, as seen in the mechanical characterization of the Historic Center of Bragança (Portugal), NDT were performed with recurrence to the application of flat monkey test presented by [3]. In another hand, the work developed by [4] in a controlled environment showed that the employment of thermography can provide a significant contribution to structural assessment, but needs more developments on this technique before be largely use by the technical community.

According to [5], non-destructive tests can be also very useful for determining the characteristics of historic masonry. By correlation between sonic waves velocities and mechanical properties from samples experimentally tested in the laboratory, it is possible estimating mechanical parameters of the structure in situ. However, this experimental was limited to direct ultrasonic measurements and was baes on ultrasonic velocities values interpretation. But, even with these limitations, ultrasonic tests stand out as quite promising NDT.

In addition, ultrasonic data has been extensively used for masonries characterization, as the works developed by [6–8], but also limited on interpretation of ultrasonic velocities pulse. Although promising, the use of more sophisticated statistical methods in the analysis of ultrasonic data to assess the condition of a building has not been widely explored, as well the interpretation of the pattern distribution of ultrasonic waves over an area are not commonly reported on the literature.

The employment of sonic tests as a characterization tool in traditional buildings has been each more reported in the literature especially since from 90 years. Although technical reports show that different types of masonries were tested, the methodology of ultrasonic data acquisition or processing has not been significantly changed since then [9–15]. Basically, the common recommendations noted of the literature consists in the necessity to perform NDT in each analyzed material and the interpretation of the results are limited to the tested sample. Basically, since from 90 decade, the waves propagation velocity values have been used as a link to correlate the properties of the assessed material.

In a report, [16] performed sonic test on the existing walls of the Cathedral of Noto. A relevant portion of this structure was damaged in 1996 due to a collapse. The intention of these investigations was to verify the state of conservation of the walls and pillars in view of the reconstruction of the damaged part of the Cathedral. The results show that sonic tests are sensible and can be successfully applied for the diagnosis and control of the effectiveness of repair techniques.

Ultrasonic test is based on the emission and reception of ultrasonic waves. The characteristic parameters of sonic waves are velocity, amplitude, phase, frequency and attenuation. The stiffness of the structure and the material density interferes with velocity, while the wave attenuation parameter is an indicator of fractures and compaction conditions [17, 18]. The levels of material deformation associated with sonic tests are sufficiently small, assuming material behavior according to the theory of elasticity as a theoretical basis for the interpretation of the results. These tests result in specific values for the velocity of wave propagation, from which the elastic modulus can be derived. Thus, making possible to characterize the material or identify intrinsic anomalies. According to [19], this method is widely used and consolidated for concrete and steel samples and consisting in emitting an ultrasonic wave through the element and measuring its propagation time.

Recent advances on employment of ultrasonic techniques for masonry characterization have point to:

- Sonic and ultrasonic tests can be used to assess the effect of retrofitting techniques in masonries, as showed by [16];
- Sonic and ultrasonic tests can used in a combined way with others NDT and semi-destructives strategies for estimating anisotropic material mechanical properties, as presented by [5, 10];
- In sonic and ultrasonic tests, the masonry joints can influence the waves behavior [20, 21];
- Ultrasonic tests can be used to identify heterogeneities in ancient masonries and to identify if a join of walls presents similar constructive properties, as well the work also shows that the loading and opening sections can influence the ultrasonic waves behavior in masonries [6];
- The presence of a mortar recovery layer (15 mm of thickness) does not presents significant influences on the ultrasonic wave behavior, as suggested by [21, 22];

In 2020, an important contribution to employment of ultrasonic test to characterize ancient masonries [23], especially concerning clay bricks masonries was the publication of the ABNT NBR 16805 [24]. In this code, a methodology for collect ultrasonic data is presented and can be used to characterize the wall homogeneity, detection of anomalies, monitoring materials changes, as well to assess retrofitting measures impact. Nonetheless, in order to enlarge the application of ultrasonic tests, it is needed to proceed with scientific investigation on how different situations or materials conditions can affect the behavior of the waves inside a material.

This way, in order to give a contribution to the advance for the employment of non-destructive tests for masonries characterization, the present work has as the main aim to propose a novelty approach for detection of voids in ancient masonries based on ultrasonic data collected by indirect measurements.

2 Statistical Approaches for Ultrasonic Data Exploration

In this section, the methodology for ultrasonic data analysis proposed in this work based on statistical tools is presented.

The frequency of a wave propagation is related with Dynamic Modulus of Elasticity. Thus, ultrasonic waves can be used to analyze material properties. In the literature, several examples of the application of ultrasonic waves for analysis and properties estimating of steel, wood, stone and concrete are often reported. According to ABNT NBR 8802 [25], ultrasonic tests performed in concrete specimens can be used to analyze the material homogeneity, estimating the compressive strength, detection of anomalies or crack depth. So, if the properties of materials are knowing and the behavior of the wave propagation across the material sections as well, this kind of testing can detect heterogeneities and be useful for estimating mechanical characteristic. Thus, characterize and understand how some anomalies interferes in the wave behavior in a determined material is the way for advances the employment of ultrasonic tests to most heterogeneous materials, as masonries, for instance.

The method proposed in this work for ultrasonic data processing initially consists into plot the ultrasonic waves data of each one frame as function of the time of wave propagation versus distance. The Fig. 1 shows the 24 curves (12 curves from P1 and 12 curves from P2), here so-called "profile" of the wave behavior. In statistic, these types of curves are often used to analyze the behavior of a data collected by repeated measures at the time.

In this work, the repeated measure was also performed. The data refer to the time that an ultrasonic wave takes to go from a point, x_0, till the positions x_1, x_2,... x_6 in each frame. Due to the methodology of data collection and considering that the material presents similar properties, it is expected a linear tendency for each one of the profiles. Based on that, some issues were highlighted in this work:

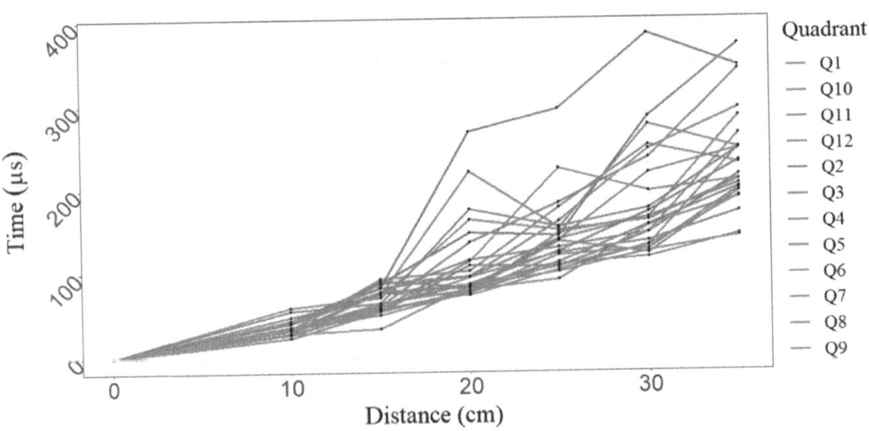

Fig. 1 Profiles of the walls frame plotted in time of wave propagation versus distance

1. Is it possible to observe any change in the linear tendency of the ultrasonic walls profiles?
2. Assuming that is possible to observe changes in the linear tendency of the ultrasonic wall profiles, can the profiles be used to identify an anomaly in a masonry wall?

2.1 Mixed Model Theoretical Background

A regression analysis with mixed-effect is a methodology where the effect of each one of the frames are considered in the analysis by including a vector of random variable in a regression model, here assigned as by **b**. This model is called in the statistical literature as a mixed model, for details about this kind of statistical model see for example [26–28].

For a brief explanation of the mixed model, consider a sequence (T_{ij}, D_{ij}) $(i = 1, 2, \ldots, N$ e $j = 1, 2, \ldots, 6)$ where T_{ij} represents the time that the wave takes to range the j-th distance, D_{ij}, in the i-th frame, and N the total of aniseed frame. Each D_{ij} is determined by the distance between the point x_0 and each point x_i, for $i = 1, \ldots, 6$, as described before, all points aligned in the same horizontal line.

The measurements of a cross of the frames is assumed as independents, and a mixed model is used to take into a count just the correlation between the data measured in each frame. So, consider that the time can be described as a function of the distance ranged by the ultrasonic wave, as:

$$T_{ij} = f\left(D_{ij}; \boldsymbol{\beta}; \mathbf{b}\right) + \epsilon_{ij}, \tag{1}$$

where f(.) can be a linear or non-linear function both in the parameters vector $\boldsymbol{\beta}$ and random-effect **b**, as the distance that is used to explain the time denoted by D; and ϵ_{ij} is the term of error, considered here following a normal distribution with mean zero and variance constant, it is commonly denoted by: $\epsilon_{ij} \sim N(0, \sigma^2)$.

The function $f(D_{ij}, \boldsymbol{\beta}, \mathbf{b})$ in Eq. (1) describes the relationship between the wave propagation time and the distance range by the ultrasonic wave for this time. This relationship can be described by a non-linear function in D_{ij}, as for instance, a polynomial curve with degree G, $G > 1$. The quantity $\boldsymbol{\beta}$ is a parameter vector that represents the common effect to all frames, this is called fixed effect parameter, and the vector **b** represents the effect of each frame in the model, and can be different across all of them, this is called mixed-effect variable.

In order to complete the model described in Eq. (1) let consider a normal distribution for each b_i in $\mathbf{b} = (\mathbf{b}_1, \mathbf{b}_2, \ldots, \mathbf{b}_N)$ as follows:

$$T_{ij} = f\left(D_{ij}; \beta; \mathbf{b}\right) + \epsilon_{ij}$$

$$\epsilon_{ij} \sim N\left(0, \sigma^2\right)$$

$$b_i \sim N(0; \Sigma_b), \text{ for } i = 1, \ldots, N, \tag{2}$$

where $b_i \sim N(0; \Sigma_b)$ represents that the bi follows a normal distribution with mean (or mean vector) μ_{ij} and variance (or variance matrix) Σ_b.

In this work a polynomial mixed effect model with $G = 3$ was choose in Eq. (2) for the function $f(D_{ij};.)$ and the vector **b** was considered as a scalar, $\mathbf{b} = b_0$. Then, this model became:

$$T_{ij} = b_0 + \beta_0 + \beta_1 D_{ij} + \beta_2 D_{ij}^2 + \beta_3 D_{ij}^3 + \epsilon_{ij}$$

$$\epsilon_{ij} \sim N\left(0, \sigma^2\right)$$

$$b_o \sim N(0; \Sigma_b), \tag{3}$$

Note that a polynomial curve can be also considered for the random-effect, in this case, the function f(.) became $f(.) = b_0 + \beta_0 + (\beta_1 + b_1)D_{ij} + (\beta_2 + b_2)D_{ij}^2 + (\beta_3 + b_3)D_{ij}^3 + \epsilon_{ij}$.

To avoid the unwanted correlation between the transformation of the distance used in the model D, D^2 e D^3 in Eq. (3), the orthogonal polynomial is used in the expression $T_{ij} = b_0 + \beta_0 + \beta_1 D_{ij} + \beta_2 D_{ij}^2 + \beta_3 D_{ij}^3 + \varepsilon_{ij}$ instead the original polynomial.

Some approaches can be used to obtain estimative of the β and b_q. In this work, the Bayesian methodology is adopted for allow also to obtain estimative of the values of b_i [29].

The disadvantage into adopt a polynomial function for modelling the global tendency of wave propagation is to impose a behavior that not necessarily occur. So, for an initial exploration of this behavior, be important to consider a most flexible method. This way, a powerful method for modelling tendencies without to impose an analytical function is presented in the next section.

2.2 LOWESS Method Theoretical Background

The LOWESS method was proposed by Cleveland et al. (2017), and initially was called Loess Method. In this approach, for each considered distance D a linear regression is fitted in the neighborhood of the D using the last square method. For an explanation of this method, consider for each considered distance D a linear regression model is fitted in a set of point in the neighborhood of D determined by a constant h as:

$$T_i(D) = \beta_0(D) + \beta_1(D)D_i + \varepsilon_i; \text{ for } D_i \in [D - h; D + h], \tag{4}$$

where T is the time expended by the wave to range the i-th distance D_i and ϵ_i is the term of error in the regression assumed with mean zero and variance constant for all ϵ_i. The points surrounding to D in the data can be choose subjectively or in a subjective form from the observed data. Note that, in this approach, several linear regressions can be fitted in the data considering only two parameter ($\beta_0(D)$ e $\beta_1(D)$). Thus, the set of the points considered in each regression can be seen as a window around each D, in the cartesian plan where each regression model is considered.

For reducing the weight of the contribution of a particular point D_i far away from D, in comparison with the other points in the neighborhood, weights can be associated with local regression models, given by Eq. (4). With this inclusion, the methodology consists basically to estimate β_0; β_1 and σ^2 in each linear regression, in order to minimize the sum:

$$\Sigma\{\omega_D(D_i)\,[T_i - \beta_0(D) - \beta 1(D) D_i]2\};\ \text{for } D_i \in [D - h; D + h], \tag{5}$$

For defining the weights ω_D (D_i) in Eq. (5), a function Kernel is assumed as follow

$$K_D(u) = (1 - |u|3)3,\ \text{for} - 1 \leq u \leq 1; \tag{6}$$

where the weights ω_D (Di) computed as:

$$\omega_D(D_i) = KD(D_i/(D + h)). \tag{7}$$

For use this model in ultrasonic data, the R package as used by the library "ggplot2" performed by [30].

3 Experimental Setup

Two specimens of typical ancient walls from Ceará State were built in the Laboratory of Rehabilitation and Buildings Durability of the Federal University of Ceará. The first wall, here assigned as P1, was built without any anomaly and take as the reference wall, while the second wall, here assigned as P2, was built with some anomalies in its interior, essentially voids. Figure 2 shows P1 and P2.

Each one of the walls were built in the dimensions of 1.50 m of high, 1.00 m of width, and 0.135 m of stiffness. To the walls building ancient clay bricks with dimensions of 4.60 cm × 24.70 cm × 12.30 cm and specific mass of 2000 kg/m^3, joined by a lime mortar in the proportion 1:1 (lime: fine aggregate, in mass) was used. The walls had one of the surfaces recovered by a 15 mm of lime mortar in the proportion 1:6 (lime: fine aggregate, in mass). The walls were left 40 days before starting the experimental measurements.

To ultrasonic data collecting the equipment model Pundit® 200 PROCEQ, with transducers of 54 kHz was used. The indirect ultrasonic method was used to collect the ultrasonic data of the walls. In this method, both transductors, emitter (E) and

Fig. 2 Ancient specimens of walls used in this experimental setup

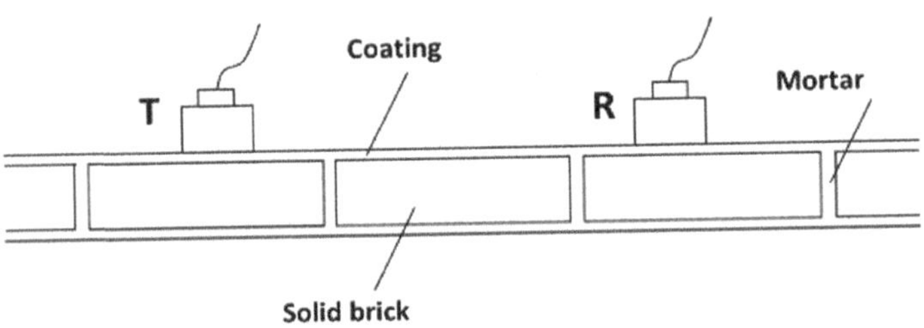

Fig. 3 Scheme of the indirect ultrasonic measurements

receiver (R) are positioned aligned in the same surface, as shows the schemes of the Fig. 3. Indirect ultrasonic method is used in the situation that the direct ultrasonic measurements are not possible, as for instance, limitation by dimensions of the assessed element.

The ultrasonic measurements were performed in the recovered surface of the walls. In each one of the walls, ultrasonic collecting data were performed in the horizontal direction (axis X) and vertical direction (axis Y). For that, the walls were divided into 12 frames with 20 cm × 40 cm and the ultrasonic measures were performed for each one of the frames. In order of the be free from walls boundaries influences, a distance of 10 cm from the wall edges were not considered during frames division. Also, 20 cm from the top and the bottom region of the walls were left out. Figure 4 shows the frames assigned in the walls.

In each frame, 6 ultrasonic measurements distributed along of the frame central region were done. The emitter and receiver transductor were distant since point x_0, till the points $x_1, x_2, \ldots, x_6$. The emitter (E) is static during the ultrasonic measurements, while the receiver (R) vary according to distances linearly increasing. So, for each distance between the point x_0 and a point x_1, x_2, …, x_6, the time of ultrasonic wave

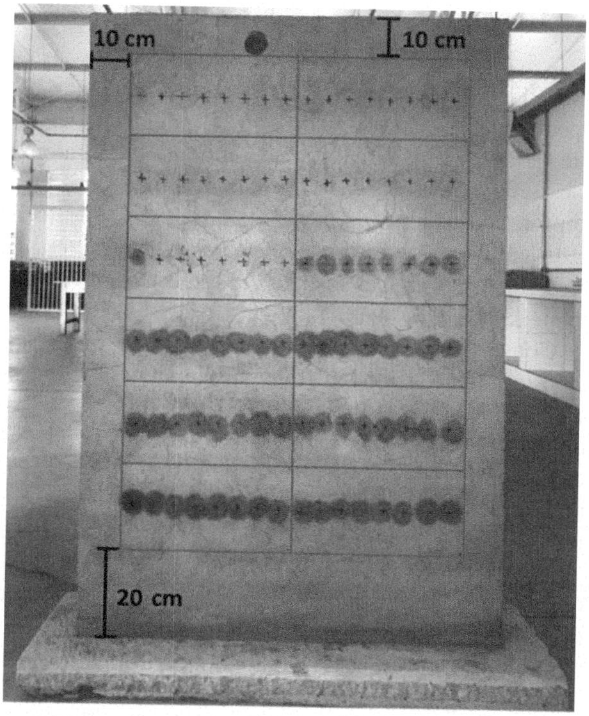

Fig. 4 Wall with the ultrasonic frames assigned

propagation was measured. Figure 5 shows the schematic view of ultrasonic indirect measures performed in the frames.

The method for indirect ultrasonic measurements adopted in this work was based in the recommendations of the ABNT NBR 16805 [24] and in the work presented by [6].

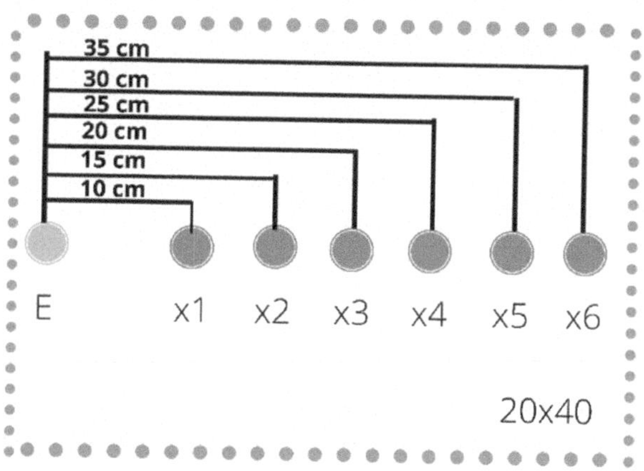

Fig. 5 Schematic view of the indirect ultrasonic measurements performed inside the walls frame

4 Results

Initially, by Fig. 1 a linear and growing tendency can be observed, once the time spends by a wave to travel from a point to other is proportional to the distance. However, by Fig. 1 analysis it is not possible to identify if all profiles are following the same expected behavior, namely, linear and growing tendency. Since a most appropriate analysis is performed, it is possible to observe that in some profiles, a disagreement with the global behavior is noted. For that, a mixed polynomial regression model was loaded and the effect of each frame to the global curve was modelled (Fig. 6).

Considering the results obtained by the mixed model, the Table 1 shows the random effect estimated from this model, showing also a credibility interval that contain the value of b_0 with 95% credibility. The credibility interval represents the influence of each frame in the global tendency of the profiles. In this table, five frames can be observed as having different behavior, with a credibility of 95%, because the random effect can be considered as different of zero with 95% of certain, since zero is included in the credibility interval. Among this frame, four of them have anomalies.

Table 1 shows the estimative of the individual frame effects to the global behavior of the curve, as well shows the credibility intervals that makes possible to identify if the frames are affecting the global behavior of the curve or not. In the case that the random effect presents zero in its interval that means that the value is very near to zero, making the distance between the obtained curve since the frame considered in relation to global curve do not be significant. Based on that, it is possible to identify anomalies on the behavior of the ultrasonic waves in the region of Q2, Q6, Q8, Q9 and Q10 from the P2 wall (Fig. 7). In addition, an anomaly as also noted in Q2 region in P1 wall. The results presented by Table 1 suggest that even if a not flexible curve

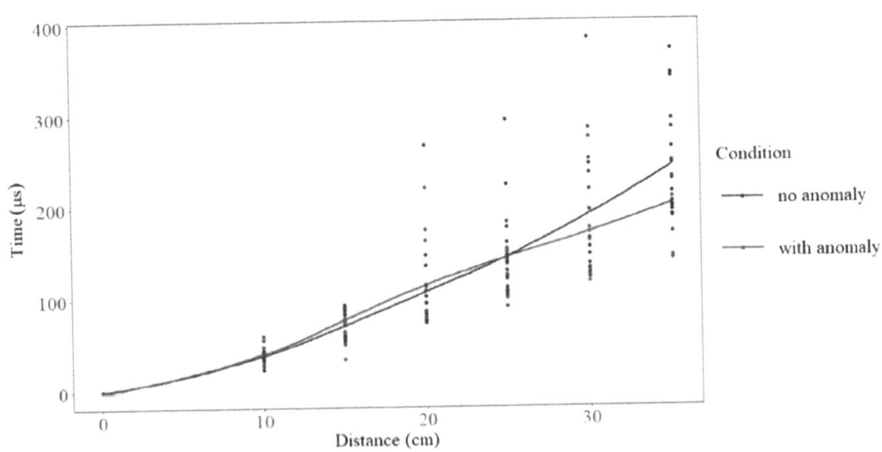

Fig. 6 Curves of the ultrasonic profiles plotted by logarithmic transform for time

Table 1 Estimated effects of the frames to estimated degree polynomial curve, G = 3, and 95% credibility intervals

Frame	Without anomalies	With anomalies				
	Average effect	Lim_{inf}	Lim_{sup}	Average effect	Lim_{inf}	Lim_{sup}
Q1	0.12	−0.05	0.29	0.13	−0.03	0.30
Q2	**0.26**	**0.09**	**0.44**	**0.47**	**0.31**	**0.65**
Q3	−0.07	−0.24	0.09	−0.07	−0.23	0.10
Q4	0.01	−0.16	0.17	0.15	−0.01	0.32
Q5	0.17	0.00	0.33	0.16	0.00	0.33
Q6	−0.17	−0.34	0.00	**−0.25**	**−0.43**	**−0.08**
Q7	−0.16	−0.33	0.01	−0.15	−0.32	0.02
Q8	−0.07	−0.24	0.09	**0.26**	**0.09**	**0.43**
Q9	0.00	−0.17	0.17	**−0.21**	**−0.38**	**−0.04**
Q10	0.05	−0.11	0.22	**−0.34**	**−0.50**	**−0.17**
Q11	−0.09	−0.26	0.08	−0.16	−0.33	0.00
Q12	−0.05	−0.22	0.12	0.05	−0.12	0.21

Fig. 7 Some panel quadrants with anomalies

be imposed, it is possible to verify frames with anomalies influencing the global tendency of the curve.

Following, the Loess model was used to analyze the ultrasonic data. For the data processing, routines R Core Team [31] were used with recurrence to ggplot2 pack available by [30] to fit smoothed curves since ultrasonic data. Figure 6 shows the smoothed curves for each one of the frames with and without anomalies.

The results have shown that the data series are largely correlated, not only because they are from the same samples, but also due to the fact that were obtained by the same points. So, it is expected that a higher correlation is observed between points from the same frame, than between different frames. Between two points observed in different frames, the wave propagation can be different according to material density.

To analyze the global tendency of the profiles, the Lowess method was used to adjust the mean curve of the data from each one wall. shows two smoothed curves where can be noted two different behavior for both global tendencies. To wall sample without anomalies (P1), it can be noted that the tendency is more linear, while in the wall with anomalies (P2), a change in the curve behavior over around the third measurement point is noted. This difference between both curves is better represented by Fig. 8.

In order to identify the curves behavior and the frames that are disturbing the smoothed curves, Fig. 9 was plotted including profiles assumed with disturbing effect to the curves. By these profiles, Q8, Q5 and Q1 presents similar behavior and with potential to influence the global tendency of the curve. To evidence the disturbing effect promoted by anomalies presence, Fig. 10 presents 3 smoothed curves, the first to the wall without voids, the second to the wall with voids presence but without the disturbers profiles and the third curve with frames classifieds as disturbers by visual analysis. It is possible to note that when the disturbers are excluded from the

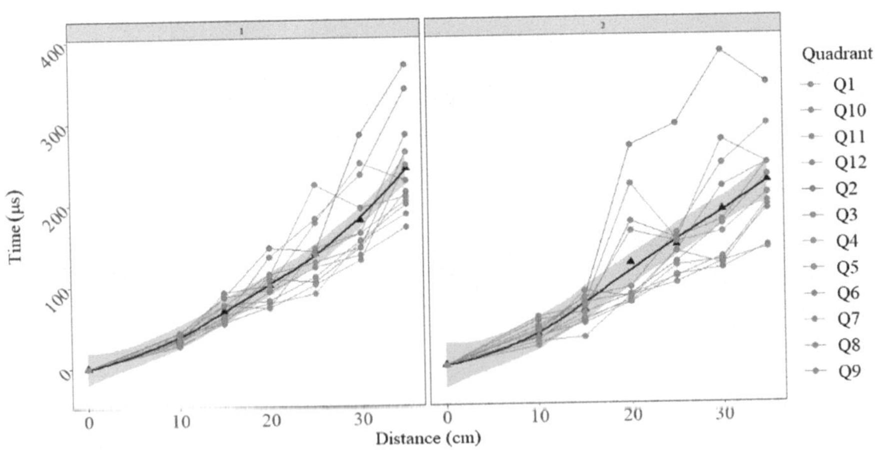

Fig. 8 Smoothed curves plotted by Lowess method

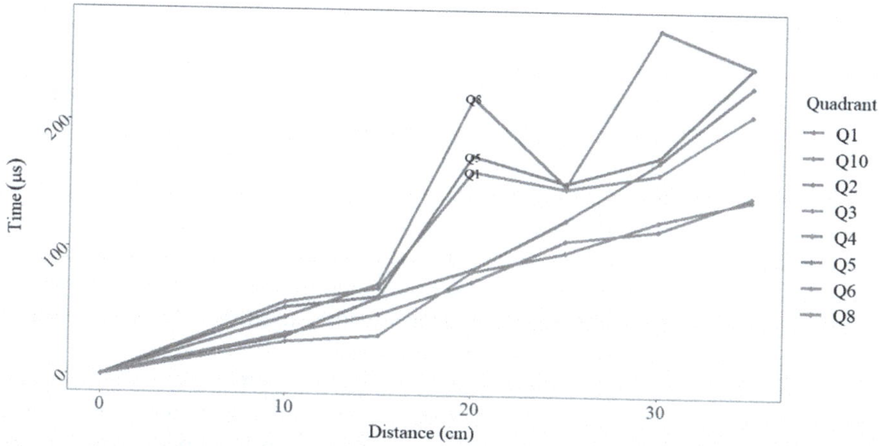

Fig. 9 Disturbers profiles identified by Lowes approach

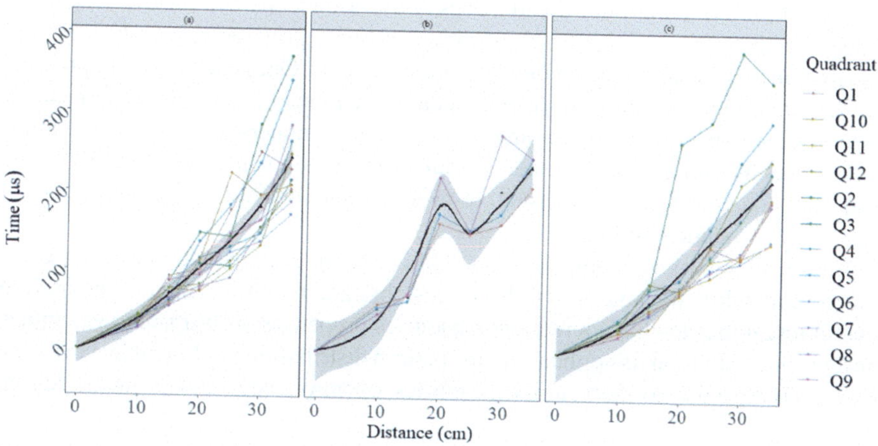

Fig. 10 Behavior of the curve profiles according with behavior

smoothed curve of the wall with the anomaly, the new curve came back to presents a linear and growing tendency.

Considering the data from ultrasonic measurements on clay bricks walls, the observation on the disturbers of the measured data shows that the voids can change the behavior of the waves in a localized way, allowing, thus, the local identification of a void even if the wall presents a cover layer with mortar, with 15 mm of thickness.

An additional advantage about the application of Mixed-effect and Lowess approaches to process ultrasonic data is the possibility of the data visualization in a panoramic way, being possible to identify how the individual measures from a frame are influencing the global data. This proceed makes possible the classification of the different behavior between the analyzed data.

5 Conclusion

Ultrasonic analysis can be a powerful tool to support materials characterization, even if this one presents a high level of anisotropy. The results presented show that the ultrasonic data can be used to analyze the presence of voids inside ancient wall structures. For anisotropic materials like masonries, the disturbances in the wave distribution profile can be used to better understand the anomalies presented as voids.

The results demonstrate how voids can influence disturb in the time ultrasonic wave propagation, making it possible to not only identify the anomalies presence, as well its location. Additionally, this method also can be used in walls with presence of a recovery layer of mortar till 15 mm of thickness.

The combined approaches employed in this work, namely Mixed effect and Lowess method, provided a novelty approach to ultrasonic data processing. While Mixed effect can be used to identify if the curves presents a data join with disturbing effect to the global behavior of the curve, Lowess method can be used to confirm the disturbers.

Thus, this work contributes to advances on ultrasonic characterization of ancient walls, introducing a new approach to ultrasonic data processing based on pattern distribution of ultrasonic data by indirect measurements. Once each masonry presents its own settlement pattern of ultrasonic data that can vary according with masonry materials composition, geometry and arrangement, to understanding how the ultrasonic waves by indirect measurements behaves is a fundamental way to analysis the voids and others anomalies presence, or even the heterogeneities between a join of walls from the same building.

Finally, this work also demonstrates that concerning ultrasonic masonries characterization, while the analysis of the values obtained by ultrasonic tests be used in several researches to characterize the masonries properties, as ultrasonic velocities, presents limitations on its application, the pattern distribution of ultrasonic waves can be an alternative way to characterize masonries, opening a new field for investigation.

References

1. E. Mesquita, P. Antunes, F. Coelho, P. André, A. Arêde, H. Varum, Global overview on advances in structural health monitoring platforms. J. Civ. Struct. Heal. Monit. **6**(3) (2016). https://doi.org/10.1007/s13349-016-0184-5
2. P.B. Lourenço, The ICOMOS methodology for conservation of cultural heritage buildings: concepts, research and application to case studies, in *Proceedings of the International Conference on Preservation, Maintenance and Rehabilitation of Historic Buildings and Structures*, 1st edn (2014), p. 12. https://doi.org/10.14575/gl/rehab2014/095
3. J.C.A. Roque, P.B. Lourenço, Caracterização Mecânica de Paredes Antigas de Alvenaria. Um Caso de Estudo no Centro Histórico de Bragança, in *Número 17* (2003)
4. S.G. Tavares, *Desenvolvimento de uma metodologia para aplicação de ensaios térmicos não destrutivos na avaliação da integridade de obras de arte* (Federal University of Minas Gerais, 2006)

5. L.U. Binda, A. Saisi, C. Tiraboschi, Investigation procedures for the diagnosis of historic masonries. Constr. Build. Mater. **14**, 199–233 (2000)
6. E. Mesquita et al., Heterogeneity detection of Portuguese-Brazilian masonries through ultrasonic velocities measurements. J. Civ. Struct. Heal. Monit. (2018). https://doi.org/10.1007/s13349-018-0312-5
7. A. Alves, *Proposição De Um Método De Caracterização De Alvenarias De Edificações Históricas Por Meio De Avaliação Ultrassônica* (Universidade Estadual Vale do Acaraú, 2017)
8. C. Maierhofer, C. Kõpp, A. Wendrich, On-site investigation tehniques for the structural evaluation of historic masomy buildings—a European research project. Struct. Anal. Hist. Constr. 313–320 (2005)
9. G. Riva, C. Bettio, C. Modena, The use of sonic wave technique for estimating the efficiency of masonry consolidation by injection, in *IB2MAC—11th International Brick and Block Masonry Conference Shanghai, China*, (October 1997), pp. 28–39
10. L. Binda, A. Saisi, L. Zanzi, Sonic tomography and flat-jack tests as complementary investigation procedures for the stone pillars of the temple of S. Nicolò l'Arena (Italy). NDT E Int. (2003). https://doi.org/10.1016/s0963-8695(02)00066-x
11. M. Guimarães, Caracterização de paredes de alvenaria de pedra por técnica sònica (Faculdade de Engenharia da Universidade do Porto, 2009)
12. M.R. Valluzzi, F.D.A. Porto, F. Casarin, N. Monteforte, A contribution to the characterization of masonry typologies by using sonic waves investigations. Résumé Keywords Masonry and building typologies (2009)
13. M.R. Valluzzi, N. Mazzon, M. Munari, F. Casarin, C. Modena, Effectiveness of injections evaluated by sonic tests on reduced scale multi-leaf masonry building subjected to seismic actions Résumé, in *Ndtce'09*, 2009, pp. 2–7
14. S.R.P.R. Matos, Caracterização de estruturas de alvenaria de pedra por recurso aos métodos do georadar, resistividade eléctrica e ensaios sónicos - Tese de mestrado (Instituto Superior de Engenharia do Porto, 2016)
15. R. D. De Ponti, L. Cantini, and L. Bolondi, "Evaluation of the masonry and timber structures of San Francisco Church in Santiago de Cuba through nondestructive diagnostic methods," *Struct. Control Heal. Monit.*, no. January, p. e2001, 2017, https://doi.org/10.1002/stc.2001
16. L. Binda, A. Saisi, C. Tiraboschi, Application of sonic tests to the diagnosis of damaged and repaired structures, vol. 34. (2001), pp. 123–138
17. J. Milsom, *Field Geophysics*, Third. 2003
18. G. Cascante, H. Najjaran, P. Crespi, Novel methodology for nondestructive evaluation of brick walls: fuzzy logic analysis of MASW tests. J. Infrastruct. Syst. **14**(2), 117–128 (2008). https://doi.org/10.1061/(ASCE)1076-0342(2008)14:2(117)
19. E.S. Júlio, Avaliação in situ da resistência à compressão do betão, in *2º Seminário - A intervenção no patrimônio. Práticas de conservação e reabilitação* (2005), pp. 41–52
20. L. Miranda, J. Guedes, J. Rio, A. Costa, *Stone Masonry Characterization Through Sonic Tests* (Porto, 2010)
21. R. Martini, Caracterização de alvenarias de granito baseada em técnicas geofísicas, mecânicas e redes neuronais (University of Porto, 2019)
22. L.F.B. Miranda, Ensaios acústicos e de macacos planos em alvenarias resistentes (Universidade do Porto, 2011)
23. E. Mesquita, R. Martini, A. Alves, P. Antunes, H. Varum, Non-destructive characterization of ancient clay brick walls by indirect ultrasonic measurements. J. Build. Eng. **19**, 2018. https://doi.org/10.1016/j.jobe.2018.05.011
24. ABNT—Associação Brasileira de Normas Técnicas, *Non-destructive Testing—Ultrasound—Panels Characterization Through Velocity of Waves Propagation* (ABNT, Brazil, 2020), p. 28
25. ABNT—Associação Brasileira de Normas Técnicas, NBR 8802:2019-Concreto endurecido-Determinação da velocidade de propagação de onda ultrassônica, 2019, p. 11
26. P. Diggle, P.J. Diggle, P. Heagerty, K.-Y. Liang, P. J. Heagerty, S. Zeger et al., *Analysis of Longitudinal Data* (Oxford University Press, 2002)

27. N.M. Laird, J.H. Ware, Random-effects models for longitudinal data. Biometrics (1982). https://doi.org/10.2307/2529876
28. M.J. Lindstrom, D.M. Bates, Nonlinear mixed effects models for repeated measures data. Biometrics (1990). https://doi.org/10.2307/2532087
29. L. D. Broemeling, *Bayesian Methods for Repeated Measures* (Chapman and Hall/CRC, 2015
30. L. Wilkinson, ggplot2: elegant graphics for data analysis by WICKHAM, H. *Biometrics* (2011). https://doi.org/10.1111/j.1541-0420.2011.01616.x
31. R. R Development core team. *R: A Language and Environment for Statistical Computing* (2011)

Pathological Manifestations in a Building at the End of Its Lifespan: A Case Study

G. A. Silva Neto, A. Tolentino Souza, C. Cavalcanti Bignoto, S. R. Souza, and W. J. Santos

Abstract Pathological manifestations are generated due to problems in the design project, building materials, execution, or use. Detection can be through visual inspection, non-destructive or destructive testing. This chapter proposes a study of the pathological manifestations existing in a building of approximately 45 years old located in the urban region of Brazil. Visually evaluated and with some non-destructive tests, all the components that compose the building (garage, entrance hall, stairs and halls of the apartments, party room and roof structure). Project management tools (three-dimensional systemic view) were used to define priorities and degree of importance of the elements and pathological manifestations present in the building. There were several pathological manifestations and flaws in the building, which were caused by errors in design project and materials, but mainly by problems associated with deficient maintenance of the building. The prioritization allowed to define an ordering of interventions and reforms by degree of urgency. The application of the solutions found for pathological manifestations aims to extend the lifespan of the building.

Keywords Pathological manifestations · Buildings · Case study · Project management tools

---

G. A. Silva Neto (✉) · A. Tolentino Souza · C. Cavalcanti Bignoto · S. R. Souza · W. J. Santos
Department of Materials Engineering and Construction, Federal University of Minas Gerais, Belo Horizonte, Brazil
e-mail: gilbertoalves.eng@gmail.com

A. Tolentino Souza
e-mail: alessandratsouza@gmail.com

C. Cavalcanti Bignoto
e-mail: camilacbignoto@gmail.com

S. R. Souza
e-mail: roberta.souzar@hotmail.com

W. J. Santos
e-mail: white.santos@demc.ufmg.br

J. M. P. Q. Delgado (ed.), *Case Studies in Building Constructions*, Building Pathology and Rehabilitation 15, https://doi.org/10.1007/978-3-030-55893-2_6

1 Introduction

Pathological manifestations in a building are due, normally, to the adoption at work of unsuitable procedures, to the non-attending of standard requirements and to the flaws in the specification of the design project and the applied materials [1]. Several international researches have shown that, on average, 10% of pathological problems in buildings originate during the phase of use of the building, in which incorrect use, mainly due to the lack of knowledge and inability of users, becomes one of the major pathological agents [2–4]. It is also observed that the quality of some buildings, specifically residential buildings, is not satisfactory [5, 6]. There is a precocious deterioration in the internal areas of the apartments, as well as in the common areas of the buildings, causing a series of expenses to the owners [6, 7].

In addition, it is possible to observe several changes made in the legislation focusing on improving the building durability. Thus, the aim is to change the current perspective in which, for instance, the degradation of concrete surfaces (320 studies in Portugal of approximately 43 years old) has a probability of 50% of reaching the end of its useful life [2]. Another recurrent pathological manifestation is humidity, which is one of the most visible problems during the inspection performed in buildings [8]. It can be conducted both in the vapor phase and in the liquid phase by diffusion, convection, capillary suction, wind pressure or gravity (water pressure) [9–12].

For the activities of evaluation and diagnosis of pathological manifestations, planning and control supported by systemic thinking are important. These tools make it possible to better conduct the execution of works and the implantation of solutions in products and services in an innovative way, which will assist in improvement of the quality [13, 14]. The fundamental benefit of this technology is the increased understanding of the system and the construction processes, facilitating the planning, project, execution and maintenance activities, and consequently improving the understanding of possible associated problems. This can facilitate the development, providing conditions for making appropriate decisions [13].

In the application of systemic thinking, the pathological manifestations in Civil Construction, associated with the various techniques, allow to analyze and solve problems caused both by poorly designed projects and by poor execution of the work. Its main techniques are: Prioritization Matrix—GUT; SWOT analysis; Action Plan, using the 5W2H tool; Flowcharts; Morphological Matrix; and Responsibility Matrix [13, 15].

This chapter consists of identifying pathological manifestations and, aided by systemic management tools, a methodology for analyzing, diagnosing and prioritizing interventions has been developed. This gathers the activities of inspection and maintenance of multifamily residential buildings, gathering a set of relevant information on the condition of the building's degradation, quantifying the effect of different degradation factors that influence the useful life of the buildings. A development plan, adjustment and maintenance strategies are presented, allowing the adoption of appropriate measures to mitigate the effects of different degradation agents.

The analyzed building was built more than 45 years ago and since its implementation, few maintenances has been carried out. Thus, the building acquired numerous problems in its structures, internal and external walls, slabs, among others, resulting from humidity and other pathological manifestations. The proposed methodology was based on an inspection of the condition of the degradation of building elements, evaluated under service conditions, using visual inspections and some non-destructive tests [16]. Qualitative and quantitative analysis of pathological manifestations were obtained based on their degree of importance.

2 The Building

The building under analysis can be classified by its typology as Multifamily Residential and in relation to the Building Inspection perspective, it can be classified as Normal (N), that is, buildings with basic facilities and equipment. It is in the city of Belo Horizonte, in the state of Minas Gerais, belonging to the southeast region of Brazil. It has approximately 45 years of existence, and its legal project was approved on August 3rd, 1975. The pedestrian and automobile entrances are directly connected to a street classified by the Brazilian system, as a local street, which means that it has low traffic volume, whose function is to enable access to neighboring lots.

The main façade faces north, and measures approximately 11.83 m of total length (railing and building) and 10.12 m of height. The building's back façade, facing south, is approximately 11.62 m long and borders an area with native forest with accentuated unevenness. The left (west façade) and right (east façade) side façades are not shared with the neighboring buildings, as they have clearances made by the pedestrian access corridor (west) and the garage access ramp (east). Each of the side façades is approximately 33.65 m long (Figs. 1, 2 and 3).

The condominium under analysis has a built area of 924.90 m^2 consisting of 3 floors and 2 undergrounds, distributed as described below:

- The underground #2 (Fig. 4) consists of a common area where we can find the condominium's point of selective waste collection, the garage, the employees toilet facilities, the cleaning material's deposit, the inactive underground water tank, the stairwell, the piped gas and the energy measurers;
- The underground #1 (Fig. 5) consists of a common area defined in project to be a party room, but unused until the time of this analysis, and the staircase. It also has a private use area of apartment 001;
- The three floors (Figs. 6 and 7) are composed, each, by a common area where the staircase is installed and two privative areas, apartments 101 and 102, 201 and 202, 301 and 302, respectively in the 1st, 2nd and 3rd floors. The areas of the odd-numbered apartments are 88.80 m^2 and of the even ones are 83.62 m^2;
- Above the roof slab of the third floor is the water tank of common use in the building, and the wood structure of the roof. They are accessible only through the trapdoor found on the ceiling of the third-floor corridors.

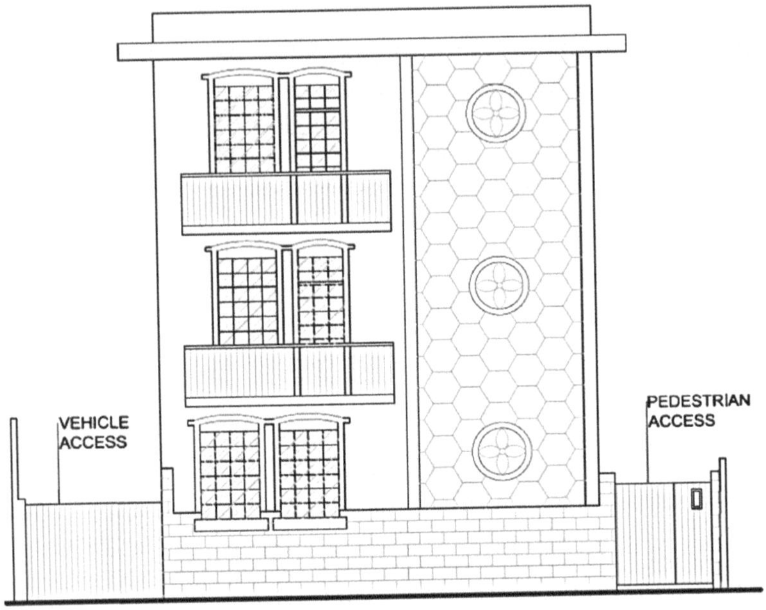

Fig. 1 North façade

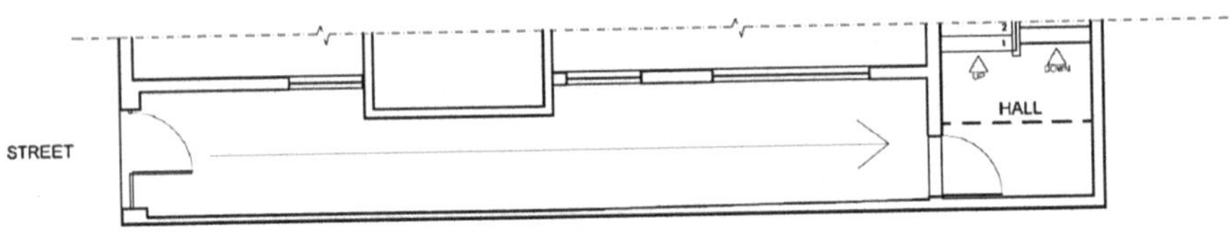

Fig. 2 Ground floor—pedestrian entrance corridor

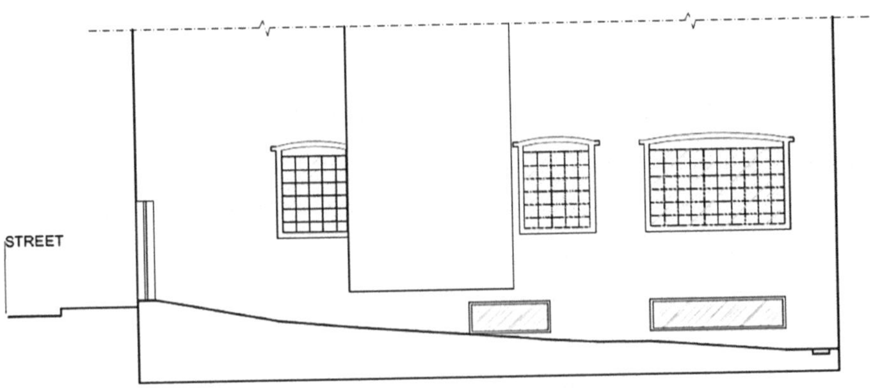

Fig. 3 Schematic elevation—pedestrian entrance corridor—west façade

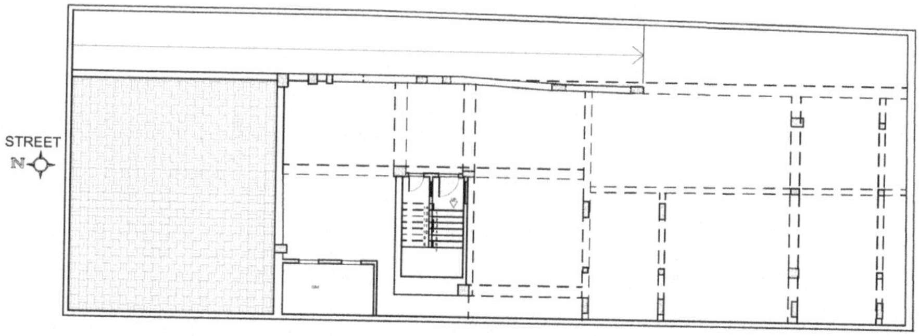

Fig. 4 Floor plant—underground #2

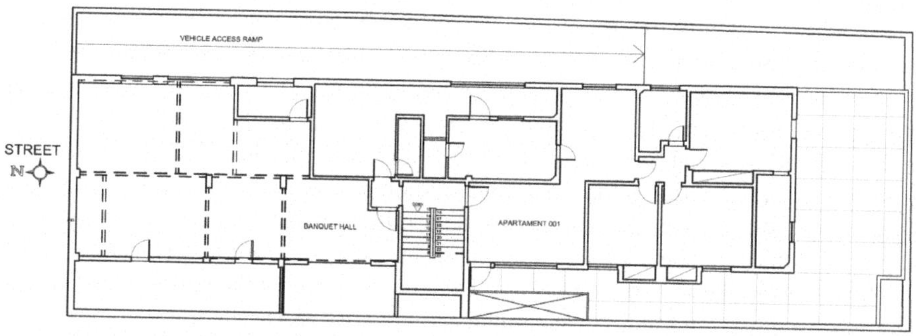

Fig. 5 Floor plan—underground #1

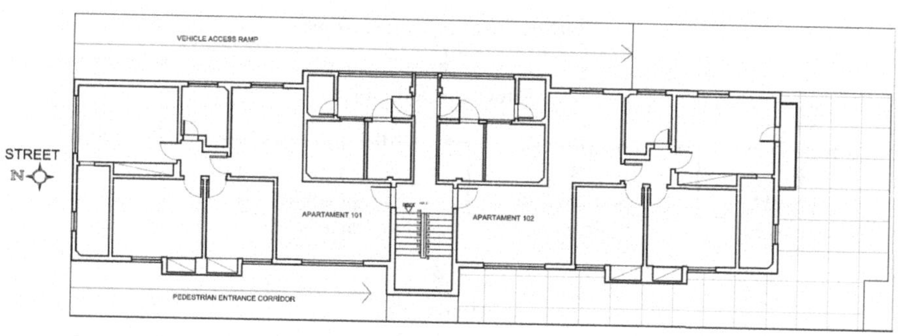

Fig. 6 Floor plan—first floor

3 Inspection Process

The inspection service was guided by the Brazilian standards IBAPE [16], NBR 5674 [17] and NBR 14037 [18], which are based on the necessary guidelines to be taken for the fulfillment of building maintenance actions with a focus on Total Quality

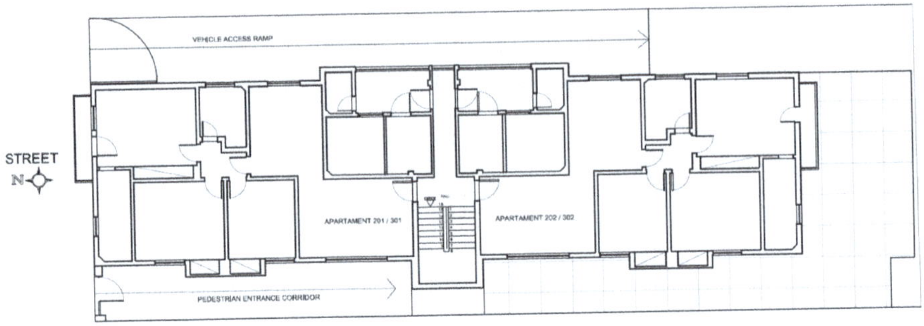

Fig. 7 Floor plant—second and third floors

(Fig. 8). This survey seeks to verify the quality of construction, maintenance and use, in compliance with operating performance, management and habitability.

The building inspection was performed based on the identification of anomalies and apparent flaws, and the classification of its level of complexity defined in accordance with the National Building Inspection Standard [16]. This process consisted of survey of subsidies through the existing documents, visual analysis at the place, and performing tests of cracks measurements, determination of powdery, and carbonation by phenolphthalein. The technical document survey was according to the annex A of the Brazilian standard NBR 14037 [18]. The classification of the building inspection [16] must be defined taking into account its complexity and preparation of the report, considering the technical characteristics of the existing building, maintenance and

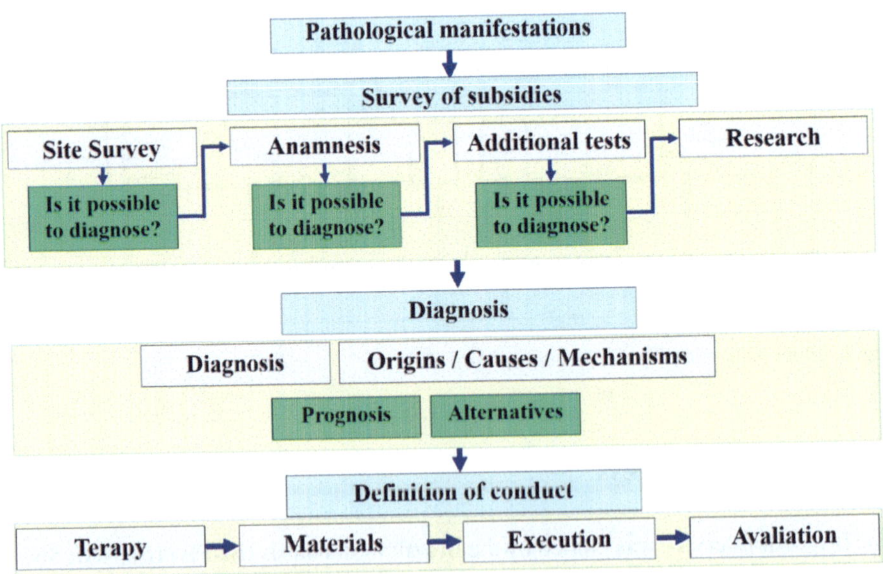

Fig. 8 Flowchart used inspection process

Table 1 Description of spaces and situation regarding the possibility of inspection

Floor	Place	Situation
2nd underground	Garages	Visited
	Staff toilet facilities	Visited
	Material storage	Visited
	Underground water tank	Visited
	Stairwell	Visited
1st underground	Party room	Visited
	Common toilet facilities	Visited
	Deposit	Visited
	Apartment 001—private area	Not visited
	Stairwell	Visited
1st floor	Apartment 101—private area	Not visited
	Apartment 102—private area	Not visited
	Stairwell	Visited
2nd floor	Apartment 201—private area	Not visited
	Apartment 202—private area	Not visited
	Stairwell	Visited
3rd floor	Apartment 301—private area	Not visited
	Apartment 302—private area	Not visited
	Stairwell	Visited
Roof	Roof	Visited[a]

[a]The visit was carried out only inside of the roof. There was no access to outside of the roof or to the water tank

operation and the need to form a multidisciplinary team to execution of the work. The levels of building inspection can be classified from level 1–3, and it is defined by the building inspector after analyzing the characteristics of the building and relating it with its purpose. The building under analysis had its inspection classified as Level 2, which, according to the regulations, is used for buildings with medium technical complexity, maintenance and operation of their construction elements and systems, medium construction pattern and with conventional systems.

The inspection covered all the common areas of the building, as well as their technical spaces, as shown in Table 1. The itinerary of inspection started with the garage, followed by the entrance hall, stairs and halls of the apartments, party room and, finally, on the roof structure. During the inspection, all the pathological manifestations and deficiencies of the building were recorded regarding the pathological aspects of the structures, services and operations. The assessment team was not allowed to access the private areas, so it was not possible to make any considerations to these places.

It was applied two methodologies aiming a technical and managerial analysis of the building conditions, being based on the three-dimensional systemic view and the prioritization of measures. The building was analyzed in terms of technique and maintenance, allowing to analyze all the variants that involve the performance of the building elements, aiming to reach the Building Total Quality. Based on the fundamentals of the GUT tools [13–15], it was possible to structure a building inspection worksheet summarizing the information contained in this chapter, which may serve as a subsidy for the services planning, the necessary works and deeper studies, if it is necessary later. This inspection defines recommendations regarding the building conditions, and it guide the maintenance planning in the condominium through the definition of the action priority based on risks, severity and urgency.

In this spreadsheet, each element was evaluated according to 9 criteria (Project, Sustainability, Preventive Maintenance, Corrective Maintenance, Integrity, Performance, Security, Administrative Management and Financial Management) classified in Lower (L), Regular (R) or Superior (S), according to Tables 2, 3, 4 and 5. Unlike traditional analysis, where the analysis is represented only by the construction, maintenance and use conditions, in the three-dimensional systemic view incorporated into the inspection process the assessment by the management criteria [13–15, 18]. This measure was introduced due to the importance of the application of process management in the management of the condominium, in order to obtain the full functioning and use of the building. Each element of the building has differentiated importance in the evaluation of the building. The functioning of the components of the firefighting system, for example, is undoubtedly more important for the integrity of the building and the safety of the users than the painting of the walls.

In order to rank comparatively the importance of each element of the building, the Degree of Importance was introduced, in which each element receives a score varying on a scale from 1 to 10. The score of each element was defined by the impact on the functioning of the building or impact on user activity, and on risk to building integrity or user security. Table 6 presents the Degree of Importance.

Table 2 Classification by technical criteria

Criteria	Project	Sustainability
Lower	Does not exist	Does not meet sustainability criteria
Regular	Incomplete, outdated	Meets sustainability criteria, mentioned in old and outdated standards
Superior	Complete, updated	Meets current sustainability criteria

Table 3 Classification by maintenance criteria

Criteria	Preventive	Corrective
Lower	Does not exist	High frequency, long response time. Never performed
Regular	Not effective	Medium frequency, average response time
Superior	Effective	Low frequency, short response time

Table 4 Classification by management criteria

Criteria	Finance	Administrative
Lower	Does not exist	Does not exist
	There is no control over the condominium's current account	There is no control over the projects. There is no control over personnel documents
	No reserve funds	There is no control over the records of condominium's protocol
Regular	Not effective	Not effective
	There is partial control	Outdated Projects
	Reserve fund incompatible with the needs of the condominium	Document control is not standardized. Record of protocols is partial
Superior	Effective	Effective.
	The current account is controlled. Reserve fund meets the needs of the condominium	Condominium has all the building projects. Personnel management is standardized. Record of protocols

Table 5 Classification by use criteria

Criteria	Integrity	Performance	Performance
Lower	Exhausted service life	Not functional	Serious, imminent risk
	Malfunctions that make repair impossible		Physical and material damage
Regular	Repairable breakdowns	Partially functional	Light risk
	Do not interfere with habitability		Long-term damage
Superior	Compatible with lifetime installation time	Partially functional	There is no apparent risk

Table 6 Importance degree of the building elements

System	Degree
Structural elements	10
Fire Fighting systems	10
Stairs and handrail	10
Financial administrative management	9
Garage operation and maneuver	9
Roof and roof slab	9
Waterproofing	9
Hydraulic, rainwater and sewage installations	9
Electrical installations, telephone and antennas	9
Sealing elements (walls and frames)	7
Painting	2

Table 7 Criticality × Weights

Degree	Gravity	Weights
Total	Loss of human lives, the environment or the building itself	10
High	Personal injury, damage to the environment or to the building	8
Average	Discomfort, deterioration of the environment or of the building	6
Low	Small annoyances	3
None	–	1
Degree	Urgency	Weights
Total	Occurring event	10
High	Event about to occur	8
Average	Event predicted soon	6
Low	Event predicted soon	3
None	Unforeseen event	1
Degree	Tendency	Weights
Total	Immediate evolution	10
High	Short-term developments	8
Average	Evolution in the medium term	6
Low	Long-term developments	3
None	It will not evolve	1

The Inspection Worksheet is the document that records the evaluation of the building systems. The evaluation depends on the level of inspection adopted. It may be visual, through documents or tests.

The prioritization of services can be done through the technique, Gravity × Urgency × Tendency (GUT) System [13–15] according to Table 7. Based on the GUT table, developed from the inspection, guidelines that will serve of subsidy for the planning and management of the condominium were determined. The method consists of scoring non-conformities as a way of obtaining priorities. Initially, weights are assigned for each non-conformity in the three functions: gravity, urgency and tendency (Table 7) and the score is obtained by multiplying the three weights.

4 Sequence and Results of Documentation

As from the technical documentation, required in the annex A of the Brazilian standard NBR 14037 [16], it was noticed that only some documents were found. Considering those pieces of information, Table 8 was structured, referring only to those documents that were found and available to analysis.

Table 8 Documental analysis

Documents	Initial delivery	Renovation	Renewal periodicity	Analyzed documentation
Legal Design Projects	Construction or Development Company	There is not	There is not	Analyzed
Meeting Protocols of Condo Installation	Construction or Development Company	Condominium	Ever change of the assignee	[a]
Condominium Convention	Construction or Development Company	Condominium	When necessary	[a]
Book of Meetings Protocols	Condominium	Condominium	With each change	[a]
Site Survey Project	–	–	–	Analyzed

[a]The evaluators were only informed of the existence of these items without the presentation of them. Therefore, they could not be evaluated and in view of this it was not possible to point out any considerations about them

When analyzing Table 8, there is a lack of important documents such as: structural and installation projects, fire department inspection report, use, operation and maintenance manuals, intervention and maintenance records, among others. The building has already undergone several changes; however, it has no register of these or even of the original materials that were used. In view of this, the evaluations were restricted to the observations made during the period of this inspection, considering the standards of the construction period, and current ones for defining prognoses.

5 Pathological Manifestations

5.1 *Infiltrations and Pathological Manifestations by Humidity*

5.1.1 Lack of Sealing of the Building's Roof and the Upper Water Tank

As noticed in Fig. 9, there are several apertures and broken tiles in the roof over the building. Besides, it was also observed the lack of roof flashing, and the painting was detaching due to humidity came from the roof and the water tank. It is important to highlight the great difficulty of accessing this place, which makes it difficult to suitable perform obligatory maintenance in the roof and reservoir.

Prognosis: The roof over the building presented serious problems of inclination, broken tiles, lack of flashings in the encounters of the roof with walls, and lack of a

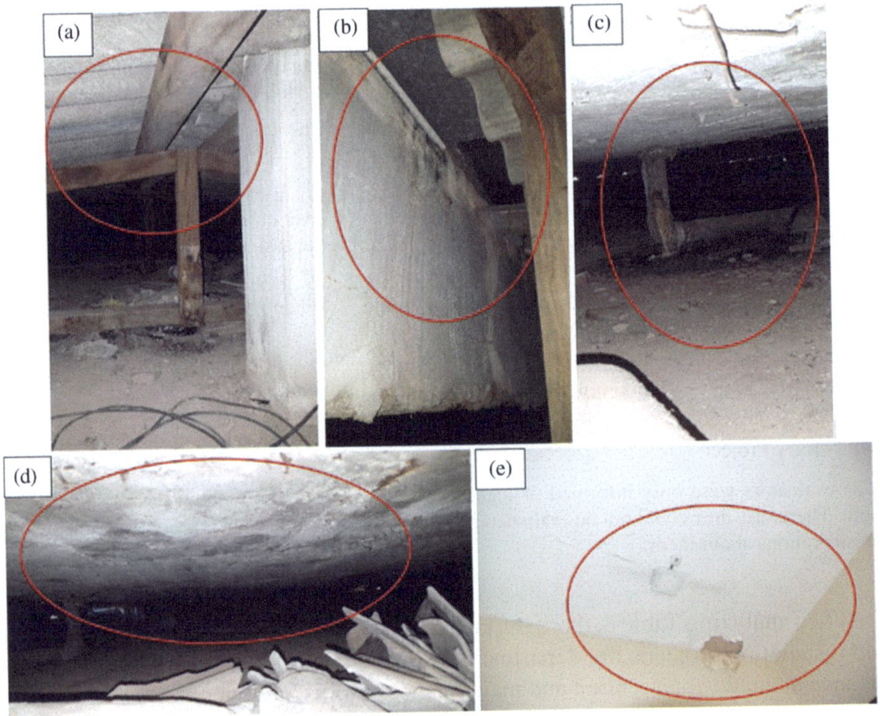

Fig. 9 Moisture from leaking roof: **a** and **b** leaking at the meeting of the roof with the reservoir wall; **c** and **d** leaking at the upper reservoir and leaking metal pipes; **e** painting peeling off in the last slab in the corridor due to the presence of moisture

maintenance/cleaning system suitable for the roof and water tank. The reservoir has some points of leakage due to problems in the waterproofing system.

Recovery Project: A qualified professional must be hired to design projects for a new roof, followed by its demolition and reconstruction, according to the legislation [19–21] and labor standards [22, 23]. It must consider the execution of an adequate structure, with materials of compatible strengths and which present security systems to guarantee the cleanliness of the roof, water tank and façades. It is also necessary to redesign and rebuild the upper reservoir [20, 21, 24].

5.1.2 Lack of Waterproofing in the Roofs of the Garage and Entrance Hall

It can be noticed by Fig. 10 stains on walls and beams due to water runoff from the roof systems in the entrance hall and from the improvised roof over the garage to cover parking spots.

Fig. 10 Moisture from leaking roof: **a** and **c** pedestrian entrance hall; **b** garage

Prognosis: This staining occurs due to the inexistence or inefficiency of flashings placed on the encounter of these walls and roofs.

Recovery Project: It is recommended to redo all the roof flashing systems in the encounter of wall and roof [20, 21] and set dates for maintenance and repairs of these systems in order to ensure the roof is watertight.

5.1.3 Moisture from Splashing on the Floor

It was observed in Fig. 11 the staining and disintegration of the painting systems and the mortar coating next to the floors of open areas of the condominium, such as the pedestrian entrance hall, access ramp to the garage and uncovered areas of the garage.

Prognosis: Staining and disintegration is due to humidity and dirt that are projected during rainy periods. These end up adhering to the wall and allow the proliferation of molds, thus the disintegration of the coating systems. This effect is maximized by the existence of old painting and mortar with high clay content.

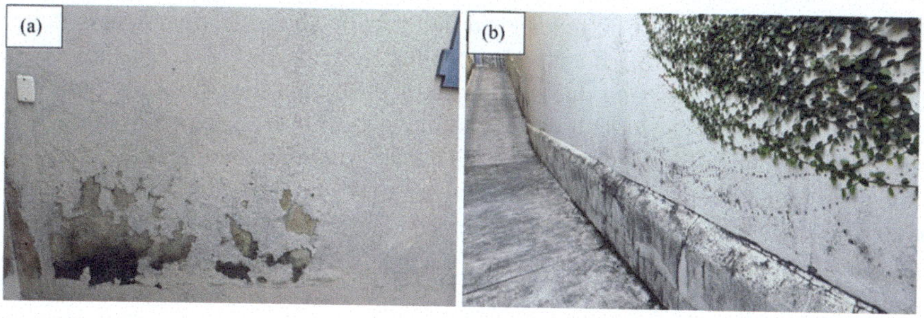

Fig. 11 Humidity from rain splashes: **a** pedestrian entrance; **b** garage ramp

Recovery Project: Places in contact with rising humidity must meet the specifications of the Brazilian standards NBR 9575 [20] and NBR 9574 [21]. All the coating system must be removed, replacing by a mortar of better quality [25]. As for painting, this system must be redone with acrylic painting in order to increase the impermeability of the wall [26]. It is necessary to put in the manual of the condominium, and effectively redo the painting system in the appropriate periods, according to the manufacturer's products and indications.

5.1.4 Moisture from Inefficient Waterproofing

As in Fig. 12, it was observed during the inspection, the presence of moisture due to cracks and inefficient waterproofing of the access ramp to the garage, as well as from the pedestrian access ramp to the entrance hall, and the open area of the private area of the 1st underground. In the case of the hall, it is highlighted the use of permeable interlocking flooring and the use of a much-deteriorated concrete on the ramp. Regarding the private area, it was found that the reform of this area was carried out less than 5 years ago (as informed by the contractor) and the waterproofing system was redone.

Prognosis: The problems were caused by errors of design project and execution, maximized by inadequate maintenance activities. In the first case, there are problems with the lack of projects and specification of waterproofing materials suitable for ramp, the common open areas and for the balcony of the private area. There are also inadequate executions regarding the performance of efficient waterproofing in these spaces, including cracks in the ramps, which promotes the percolation of water.

As for maintenance, the use of pressurized water to clean and remove sludge, maximizes the percolation of water through the cracks in the middle of the floors and in the encounters with the wall. At the intersection of the walls, a material was even placed in order to reduce water percolation, however the wrong material was used (Portland cement mortar). In this case, a movement joint must be applied together with an elastomeric sealant.

Recovery Project: It must be redesigned and rebuilt: the pedestrian access ramp, with the implementation of floor waterproofing systems [20, 21]. On the access ramp to the automobiles, and due to the need to maintain the use of the building, it is recommended to carry out a waterproofing system over the current system and to create mechanical protection over it [20, 21]. Attention should be paid to the movement joint systems at the meeting between the floor and wall of the building in order to avoid cracks and thus water percolation. The entire wall cover of the affected areas must be redone [25, 26].

Fig. 12 Moisture from inefficient waterproofing on the access ramp to the garage and pedestrians entrance: **a** in the garage; **b** in the party room—garage ramp; **c** garage access ramp; **d** detail of the meeting of the ramp floor and building wall; **e** fixing and tightness of the pipe meeting to sewage drain in the encounter with the slab structure; **f** disaggregation and proliferation of vegetation due to the presence of moisture and use of contaminated aggregate; **g** basement in the 1st underground; **h** wall of the party room in the 1st underground

5.2 *Pathological Manifestations in the Structural Elements of the Building*

5.2.1 Horizontal Cracks in the Garage's Columns

As seen in Fig. 13, it was noticed horizontal cracks in the four columns and beams at the back of the garage (supporting the loads from the open area of the apartment 001), next to the encounter of the columns with the beams.

Prognosis: Such cracks possibly come from foundation settlements. It is important to evaluate individually these cracks to conclude if there is an evolution of these aperture. These cracks were measured with the aid of a fissurometer, and it was noticed that their dimensions ranged from 0.3 to 1.8 mm. The Brazilian standard NBR 6118 [27] looking for a good performance of the structures in relation to the protection against corrosion of the reinforcement and the sensorial acceptability of the users, assigns limits to the opening of these cracks. Sensory acceptability is understood as the psychological discomfort generated in users, even if these cracks do not represent a loss of safety of the structure. As the class of environmental aggressiveness in which the building is inserted is moderate urban (CA II), the cracking requirement is maximum of 0.3 mm (according to Table 13.4 of the referred standard). Therefore, the opening in question is over the allowed limit.

Recovery Project: It is suggested to hire a specialized professional in structural design to evaluate the situation. If the evolution of the cracks in the columns is verified (active cracks), it is recommended to hire a specialized team to evaluate the case and, if necessary, reinforce the foundation to stagnate the possible existing settlement. In case of the cracks are stagnant (passive cracks), it is only necessary to close them applying epoxy resin, in order to protect the reinforcement and guarantee the psychological comfort of the users. Since the dimensions of the crack openings do

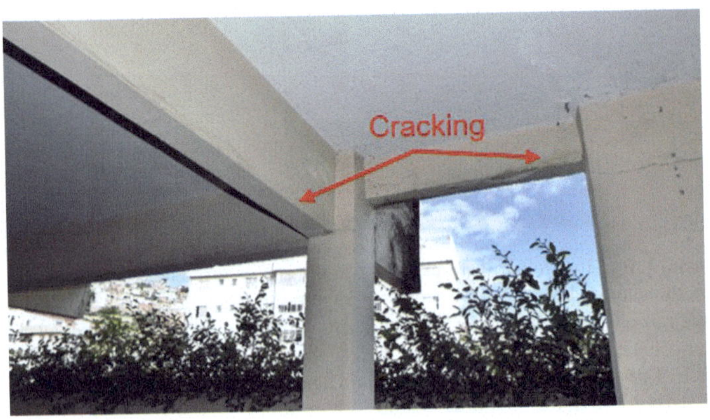

Fig. 13 Horizontal cracks in the columns at the back of the building's garage

not attend the limits [27]. The sealing problems of the sewage and rainwater systems described in the topic 5.4.2 (Problems in rainwater, sewage and water installations) may be maximizing this effect, silting the soil under the foundations.

5.2.2 Cracks Close to the Longitudinal Reinforcement and Concrete Swelling

It was detected that the intermediate columns and some beams of the underground presented cracks near the longitudinal reinforcements. These were in the middle of the beams (maximum load) and at the bottom of the columns (maximum load). In addition to the cracks, it was also detected the swelling of the concrete (Figs. 14 and 15). And it is possible to see stains around the upper reservoir.

Prognosis: Several carbonation locations were found in the structural elements (Fig. 15a, b), even detachment of the concrete (Fig. 14b), and consequently the concrete deterioration and reinforcement corrosion. It was observed that the passivation layer of the reinforcement (concrete covering) was insufficient (less than 1 cm), reaching zero centimeters in some points, which the reinforcement was facing the structural element. The region analyzed was carbonated (Fig. 15). Due to the absence of previous projects and the constructive failure detection, the reinforcements had little coverage and the structure behavior indicated the corrosion of its longitudinal reinforcement in columns, and beams in the most loaded regions. As well as at the bottom of the upper reservoir, generating the aspects identified and raising risks to the building. It is noticed that the regions that these events are already advanced, it showed humidity presence through the whole life (at the reservoir) or in rainy periods (at the garage).

Recovery Project: The base of the columns, and the beams and columns regions identified with cracks, swelling and reinforcement corrosion should be scarified, after a previous shoring of the beams and slabs in the damaged region. It is also important to make a more accurate analysis of the already compromised area and of the other structural elements where the pathology manifestation has not yet been visibly seen. Such service must be performed by a technical specialized equip in structural reinforcement, and the therapy adopted must be designed and performed by a qualified professional.

The covering and reconstitution of the currently exposed reinforcement must be carried out urgently to avoid the maximization of their corrosion and possible loss of the load bearing capacity of these structural elements. The recovery project aims to recover the support beams and columns of the building; the affected area should be demarcated, scarified by hand removing any loose or segregated material until it reaches healthy concrete, after this the corrosion must be removed replacing the reinforcement, if they are more than 10% damaged. If there is any loss of reinforcement, the complement is necessary. Some substitution techniques can be performed, in this scenario the damaged reinforcement should be removed, and another steel bar must be welded with overlap on both sides [27].

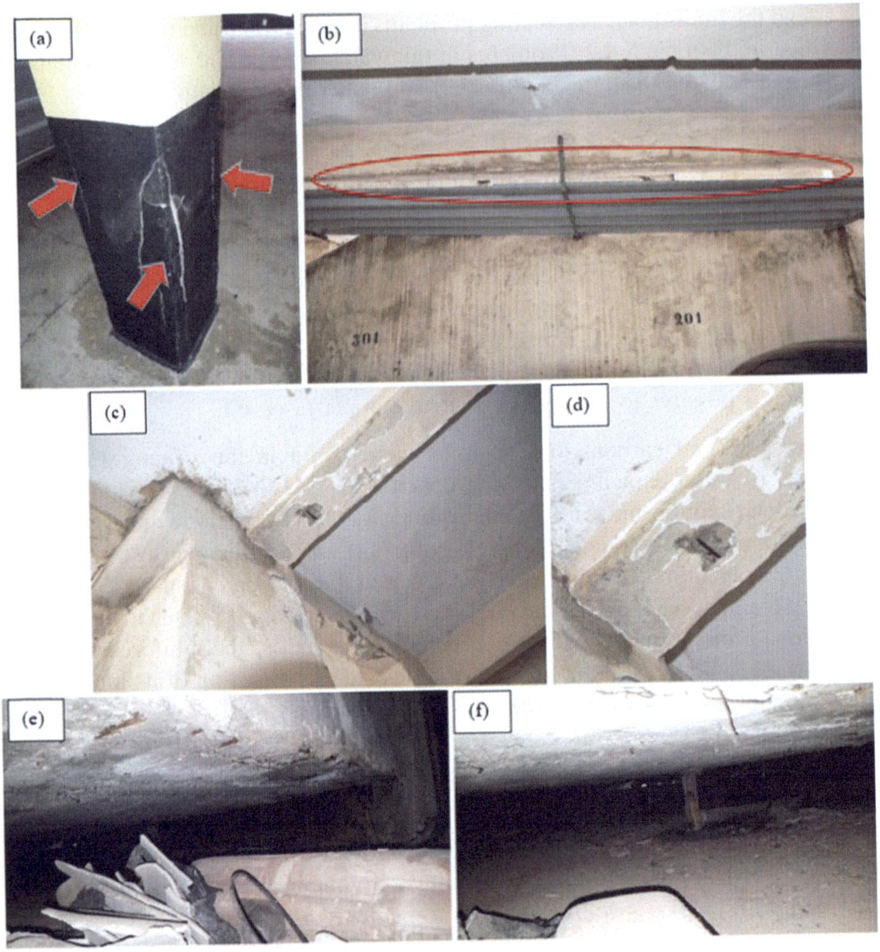

Fig. 14 Cracking and swelling of structural elements in the building's garage: **a** column; **b** beam; **c** exposed reinforcement on the beam; **d** detailing of the exposed reinforcement and **e** and **f** detailing of exposed reinforcement of the upper reservoir

In addition, it is important to clean the surfaces by water blasting with pressure and steel brush; apply "Adhesion Bridge" over old concrete, consisting of mineral or acrylic based adhesive; apply cement-based polymeric mortar; perform the final finishing; and finally, cure the surfaces by humidity, for seven days.

The best solution for constructions to be carried out is to ensure adequate coverage with the concrete properly well-dense with enough thickness to prevent or delay the penetration of deleterious materials as chlorides and sulfates. In the elements that this study did not identify problems that need corrections (after a more detailed evaluation), it is suggested to apply waterproof painting or mortar to avoid penetration of aggressive agents and spread reinforcement corrosion.

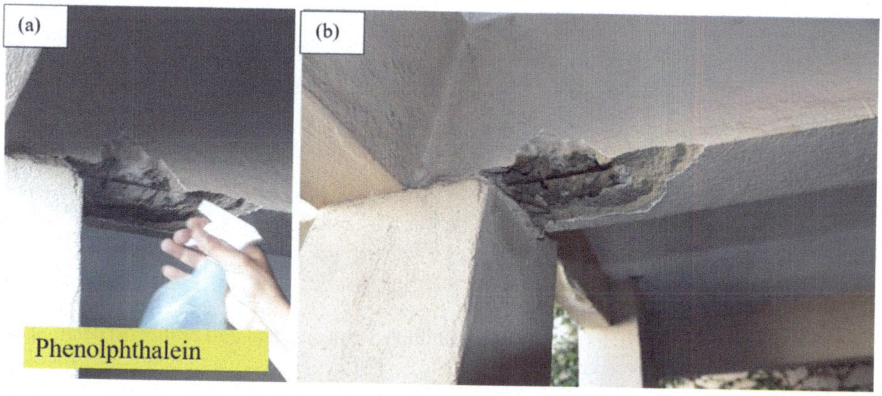

Fig. 15 Lack of covering of the reinforcement of the beams and columns of the building's garage: **a** application of Phenolphthalein and **b** carbonated concrete and without reinforcement covering

5.2.3 Cracks in the Encounter of Structures with Different Ages

Figure 16 shows vertical cracks in the region where the ancient building's structure meets a new structure done to expand/adapt the building's entrance. This crack extends throughout the whole meeting of the two different constructions (floors, slab and walls).

Prognosis: Such cracks are caused by the differential movement of each building, and as they were not efficiently consolidated, a crack appeared in this encounter region.

Recovery Project: It is suggested to hire a qualified professional to evaluate the structural part and define corrective measures through an intervention and recovery

Fig. 16 Crack on the wall (**a**) and on the floor (**b**) of the entrance hall

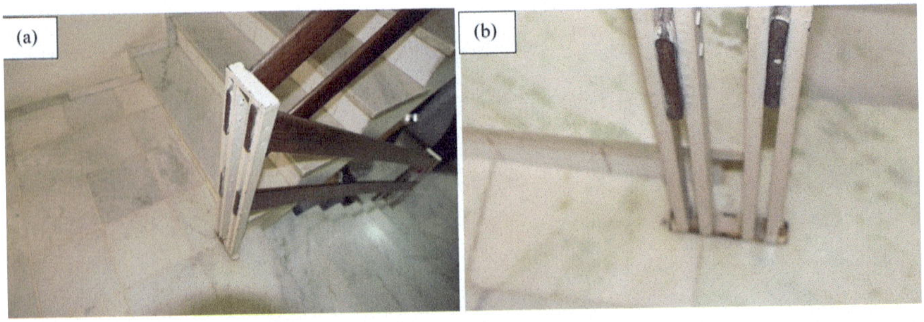

Fig. 17 Handrail of access stairs to the condominium apartments: handrail profile (**a**) detail of the fixing system to the floor presenting corrosion (**b**)

project for this part. This study highlights the urgency of these intervention to avoid water percolation and thus increase the damage.

5.2.4 Inefficient Stair Handrail

Figure 17 shows the situation of the handrails of the access stairs to the building. It should be noted that these should have guardrails and handrail, however their height is insufficient (1.20 m), in disagreement with NBR 14718 [28].

Prognosis: There is no proper guardrails and handrails, with a high risk of falling due to inefficient fixation (they are loose) and due to the corrosion process in the structure in the fastening screws.

Recovery Project: Redesigning and rebuilding the guardrails with the handrails in a proper way, according to NBR 14718 [28].

5.3 Pathological Manifestations on Building Coatings

5.3.1 Dirt Stain on Coatings

In Fig. 18a–d, it is possible to notice stains on the coating, caused by the carriage of dirt by rainwater.

Prognosis: The appearance of this staining occurs due to the inadequate design and/or execution, regarding the construction of elements to interrupt these events. Because of this, there are stains on the wall from inefficient drips under the windows (Fig. 18a); lack of drips above the boundary walls (Fig. 18b) and lack of waterproofing of the encounter of the roof and wall of the building (Fig. 18c, d).

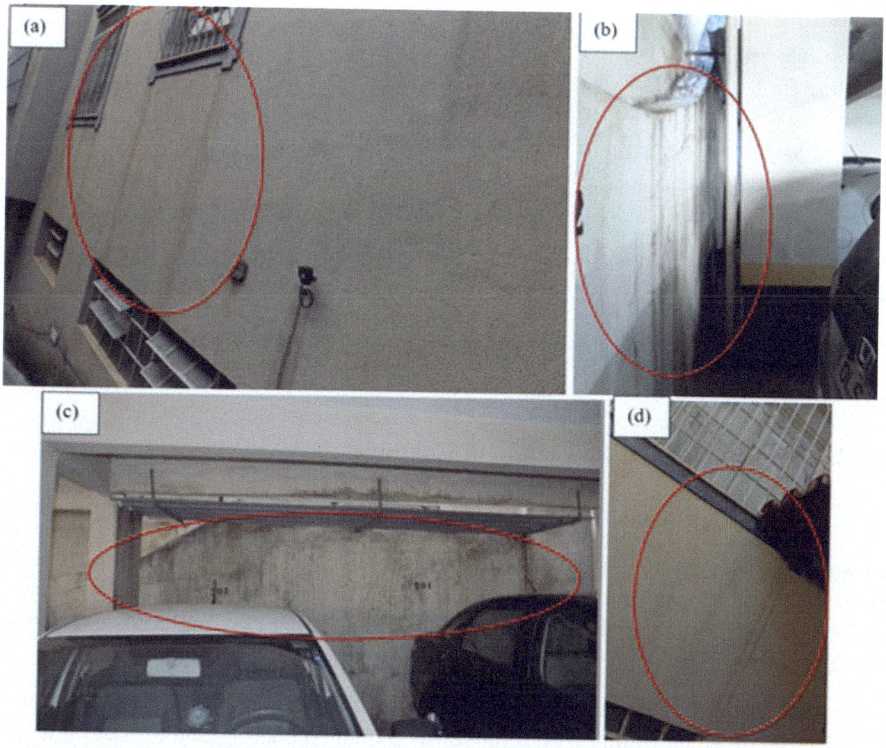

Fig. 18 Stains on the wall due to carrying dirt: **a** inefficient drips under the windows; **b** lack of drips above walls; **c** lack of waterproofing against the roof and stop of the garage and **d** lack of waterproofing of the encounter of the roof and wall of the building of the west facade of the building

Recovery Project: It is suggested to hire a qualified professional to make waterproofing projects and to carry out the execution, according to NBR 9575 [20] to select the proper waterproofing and guide the project, and NBR 9574 [21] for the proper waterproofing execution. Furthermore, it should be highlighted the importance to make drips on the boundary walls after completing the repair service, in order to minimize humidity and consequently the pathologic manifestation in study. NBR 15575 [29] demonstrates that to avoid stains close to the windows on the façade, the encounter of different levels and materials should be protected with flashings, and the windowsill must have drips, in order to avoid drainage of water through the structure.

Thus, it is noted the need to use window-sills with appropriate drips, to avoid these stains on the façade. Another factor that is causing stains is the details around the windows. It is important to waterproof the encounter of the details and the façade, or even remove it, in order to avoid points of water accumulation. After changing or adapting window-sills, it is recommended to clean and painting the façade and frames [26, 29].

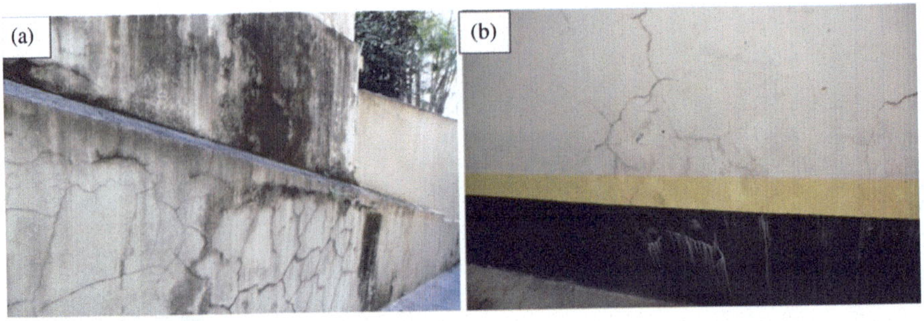

Fig. 19 Detachment of coating mortar and painting: **a** on the walls near the garage; and **b** on internal walls of the garage

5.3.2 Detachment and Retraction of Coatings

In Fig. 19a, b, there are detachment of coating mortar and several shrinkage cracks.

Prognosis: The coating is detached from the structure and it presented cracks (Fig. 19a), it is loose and detaching due to the excess of fine materials (gravel) in the coating mortars, and humidity presence generating clay materials expansion and leaching of fines, causing the appearance of efflorescence (Fig. 19b)

Recovery Project: There is no recovery for this type of pathology. The solution is to solve the problem (humidity), remove the damaged coating and execute it again correctly [25].

5.3.3 Disaggregation and Powder Behavior of Coatings

In Fig. 20a–h, it is possible to see powdery disintegration in mortars and paintings.

Prognosis: The coating is detached from the structure, loosening and disintegrating (powder behavior). Because of the coloration presented, it is suggested that a coating with the addition of grit (designation of soils from the decomposition of schist and gneiss) [31] was used, or clays were used as plasticizing materials in mortars. According to standards, such additions are used to improve certain properties of the mortar. However, the use of grit, widely used in the past to replace hydraulic lime, used in and indiscriminate manner is a major cause of pathologic manifestations such as cracks, efflorescence, powdery and detachment. Because on the contrary of the hydraulic lime, grit does not have agglomerating properties. In this part of structure, there was a large area of powdery detachment, problem which can be worsen due to moisture penetration in the structure. This is confirmed by the growth of vegetation on the damaged coat, indicating the presence of water and organic material. Organic material from the use of grit in mortar and the water is from inefficient waterproofing of apartment 001 drain, situated above, as the water descent is on this column.

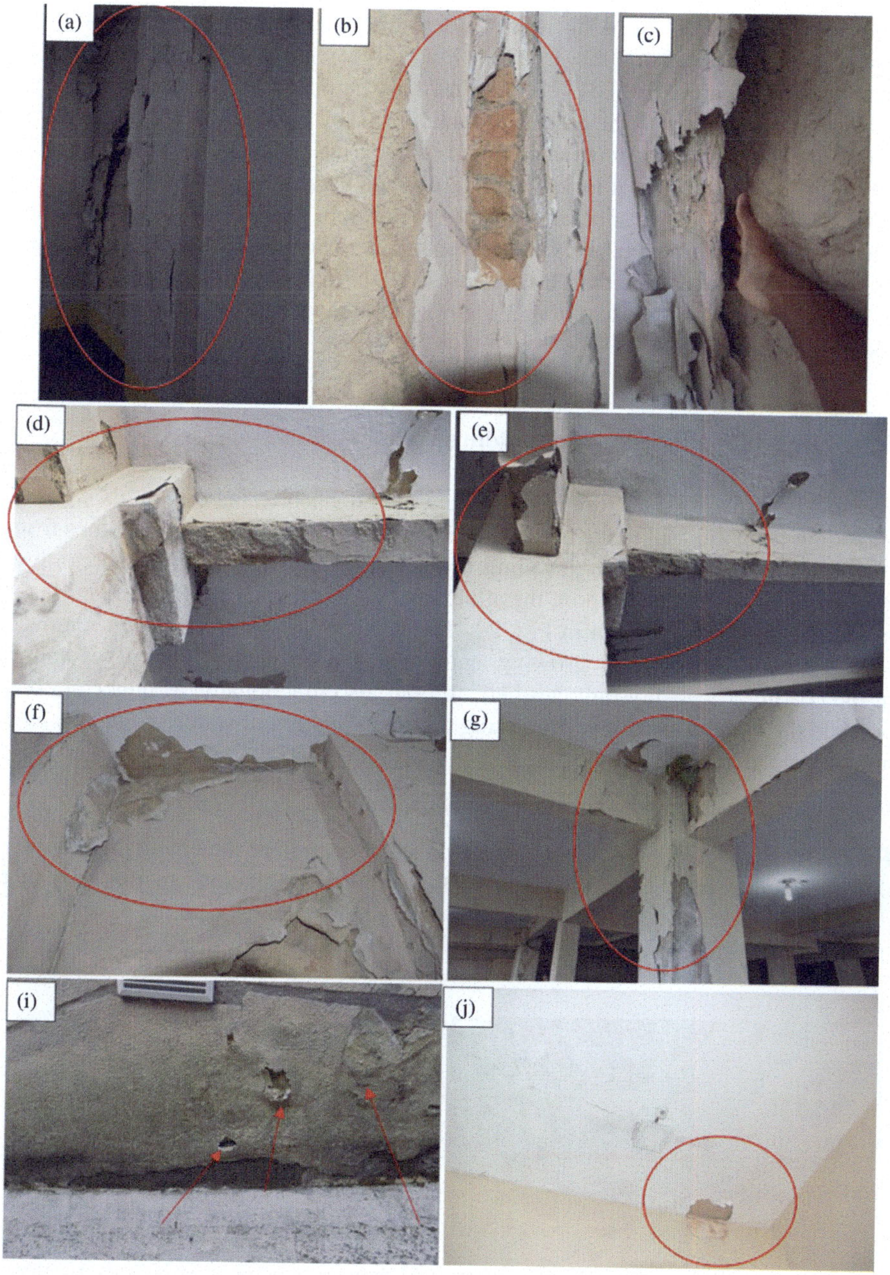

Fig. 20 Powdery disintegration of mortars and paintings: **a**–**c** on the garage walls; **d**–**g** encounter of slabs, columns and beams; **h** on the walls of the building close to the floor and **i** on the ceiling of the last floor (under the roof)

Recovery Project: There is no recovery for this type of pathology. The solution is to solve what is causing the problem (humidity), remove the damaged coat and execute it again properly [25]. The mortar must be dosed, or it is possible to use industrialized mortar with the proper care established by standard, such as pre-wetting of the substrate, proper dosage and choice of materials, and an efficient cure. NBR 7200 [25] establishes that it is necessary to execute the roughcast on the substrate in order to promote greater adherence of the mortar with the surface of the substrate. The base of application must have a certain roughness in order to increase the contact surface between layers and thereby increase system's adherence. This roughness can be produced through the execution of the roughcast, avoiding the detachment of the coating mortar.

5.3.4 Vegetation Fixed to the Building Walls

Figure 21a, b shows the presence of vegetation attached to the building's coating, spreading from the wall to the building's boundary wall. The vegetation comes from the back allotment, and it is invading the building's area, rooting in the coating. According to contractor's report, the allotment belongs to the City Hall and its vegetation is very dense, as seen in Fig. 22. The vegetation has already been removed from the walls, but it keeps growing again. The growth of this type of vegetation in the coating affects not only aesthetically but also functionally, as they can open pores in the coating structure, promoting the accumulation of water, which can lead to adherence loss.

Prognosis: The coating is damaged by the presence of vegetation.

Recovery Project: Removal of vegetation or substitution of the coating by another that is suitable for the proliferation of this type of vegetation. In the case that contractor adopts the removal of vegetation, this removal must be done periodically, according to vegetation growth.

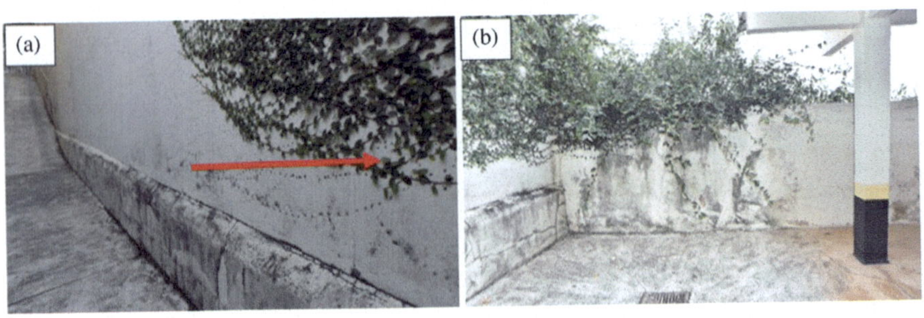

Fig. 21 Vegetation rooted in **a** the wall of access ramp to the garage and **b** boundary wall

Fig. 22 Situation of the boundary wall at the back of the building

5.3.5 Lack of Coating on Boundary Wall at the Back of the Building

Figure 22 shows that the outer part of the boundary wall of the back of the building has no protection against the weather, the rain, humidity and even vegetation can penetrate through the bricks and/or openings in the bricks, seen in Fig. 22. It is important to emphasize that excessive humidity is harmful to coating and compromises its functionality.

Prognosis: The wall and the coating are compromised by the presence of vegetation and lack of protection of the outer part of boundary wall (proper coating).

Recovery Project: Removal of vegetation or coating execution, following the procedures of ABNT NBR 7200 [25]. The materials chosen must be suitable for the proliferation of this type of vegetation, since this area is external to the building, and interventions are more complex and needs City Hall authorization.

5.3.6 Inappropriate Coating for Columns and Beams of the Garage

Figure 23 presents the situation of the coating and painting systems of one of the columns of the garage. This situation happens mainly in the columns closer to areas with great water contact.

Prognosis: It is noticed that the structural elements did not have roughcast as established by standardization [25]. The lack of roughcast impairs the adherence and can cause coating detachment. The analyzed coating easily detached from the columns, in addition it presented powdery behavior (poor quality mortar), it crumbled into powder when touched, highlighting that probably the material used was not properly dosed or it did not possess adequate binding properties.

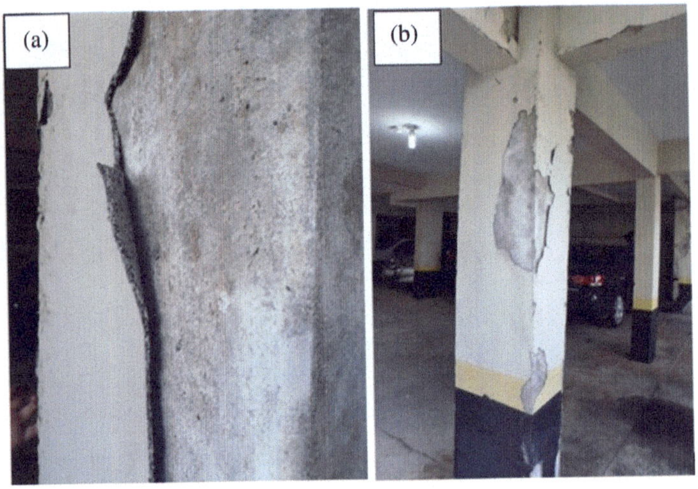

Fig. 23 Detachment of the painting system on the columns

Recovery Project: Removal, and reconstruction of the coating in an adequate manner [25], as described in the topic 5.3.3.

5.3.7 Aged Painting System

Figure 24c shows that the wall has several holes in the coating, which according to contractor's report were due to previous use of a resident to install cages for chickens. However, the cages were removed, and the holes remained. These holes affect the aesthetics and can affect the coating integrity, as they are possible access for humidity and contaminants for the structure. The stains (Fig. 24a, b and d) were due to the painting system and mortar ageing, and seasonal rain.

Prognosis: The coating and painting system were aged, and because of this there a need to reform the damaged elements on the building's façades and common areas.

Recovery Project: Removal and reconstruction of the damaged mortar [25], and reapplication of the painting system [26].

5.3.8 Lack of Lintels and Counter-Lintels in Doors and Windows

Figure 25a, b presents cracks leaving from the corner of the door at 45°.

Prognosis: the crack presented is geometrical and occurs due to the deficiency or lack of lintels and counter-lintels in doors and windows. The characteristic of this occurrence is a crack of 45°, in relation to the horizontal, at the corner of these elements.

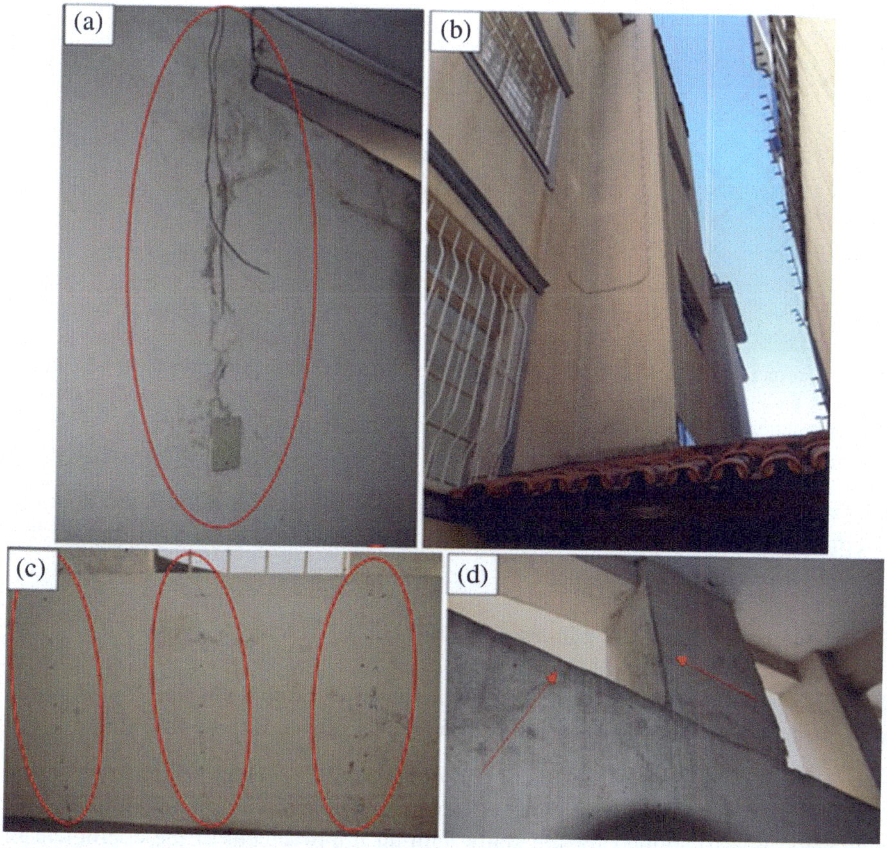

Fig. 24 Stain and abrasion by age of mortar and painting system: **a** party room; **b** façade of the building; **c** party room; **d** side opening of the garage access ramp

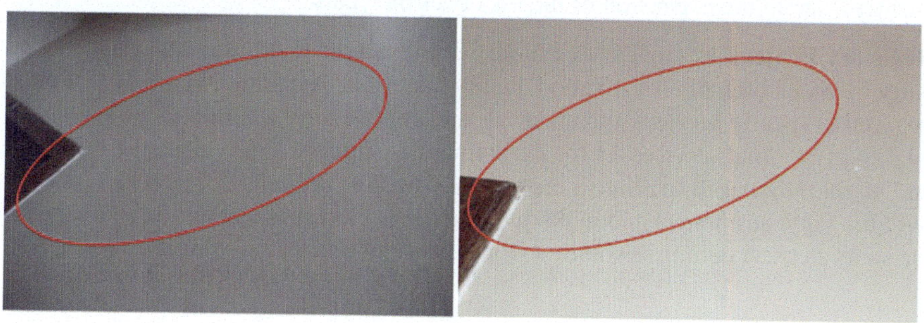

Fig. 25 Fissure at the corner of the door of top floor's apartment

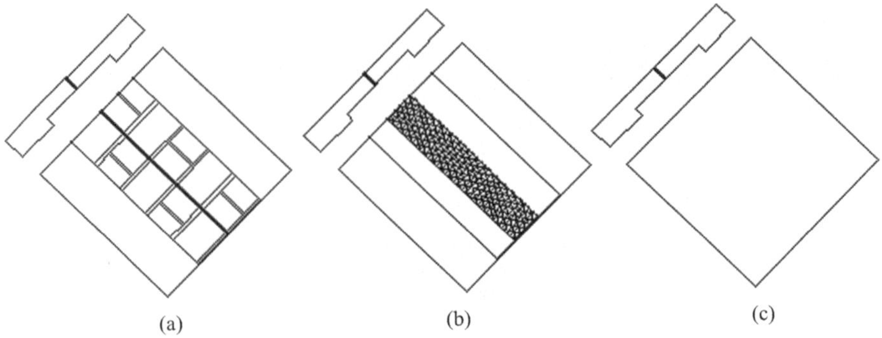

Fig. 26 Correction of cracks caused by tension at the edges of windows and doors with wire mesh: **a** opening the wall's coating 30 cm on each side; **b** fixing a wire mesh or 20 cm bandage on each side; **c** execution of the new coating

Recovery Project: Remove the mortar around the crack at windows and doors in 30 cm on each side of the crack, staple a wire mesh and remake the coating (Fig. 26).

5.3.9 Damaged Ceramic Floors and Tiles

In Fig. 27a–d, it is possible to see several damaged ceramic floors and roof tiles due to the use and inefficiency on specification and execution of these finishing.

Prognosis: Ceramic floor tiles coating were completely damaged and need to be rebuilt with another type of ceramic, or other material, more resistant to traffic. This suggestion is also applied for the stairs with an improvement of the coefficient of friction of the floor for these places that configure an escape route for fire scenario. The coating of the bathrooms was adhered with an old mortar and the detachment occurred due to the expansion of the tiles over the years.

Recovery Project: Remove the ceramic coatings of the first level, underground and party room, and rebuild it with good quality materials. A qualified professional must be hired to specify services and materials and control the execution to guarantee the quality of the final service. As for the stairs, it is suggested to replace the materials and marbles damaged (palliative method, as it will be difficult to find such elements with the same aesthetics) and making the friezes or attaching an anti-scrape tape on the stairs steps. A definitive solution would be to rebuild the building's stair [29, 32].

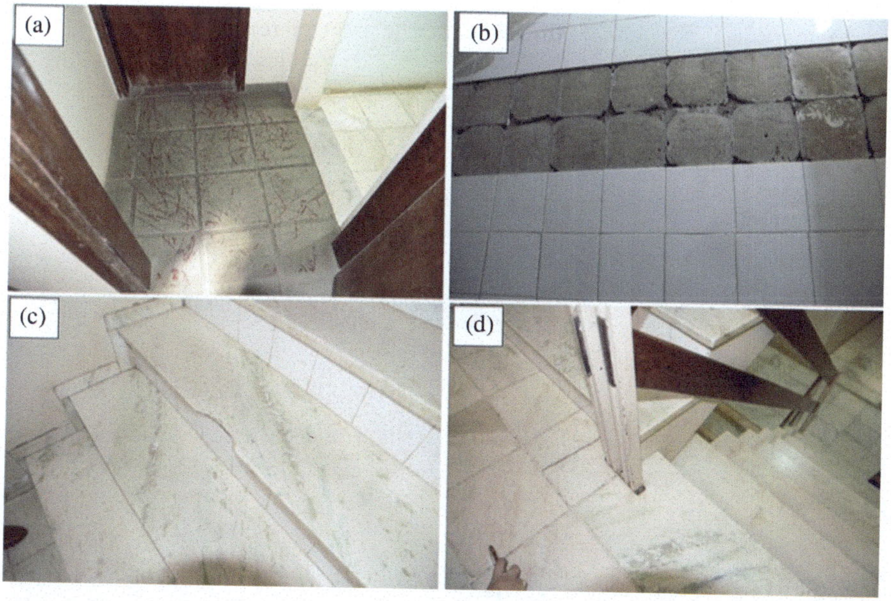

Fig. 27 Floor situation: **a** painted ceramic tile in the entrance hall to the 1st underground and party room; **b** bathroom wall coating of the garage; **c** marble floors **d** and ceramic floors of the stairs from the access hall

5.4 *Pathological Manifestations in the Building Facilities*

5.4.1 Problems in Electrical Installations, Telephony and Antennas

Figure 28a–e show several problems in electrical and telephony installation, and antennas. It is noteworthy that these were recently remade by and electrician (information from apartment manager).

Prognosis: The electrical and telephony/internet installations and antennas were not made by a qualified professional and show several disagreements with the design and execution legislation [33], generating insecurity for residents. These installations presented several improvised constructions carried out by the electrician and/or residents to meet specific demand of each one, without a precise analysis of the effects of these actions on the building.

Recovery Project: It is necessary to hire a qualified professional to evaluate the conditions of the electrical, telephony and antennas installations, promoting an efficient intervention without patches. It is believed that there may be leakage current, and the current conditions lead to risks for residents and higher energy consumption, in addition to the risk of energy loss and equipment burning due to atmospheric discharge (which there were no system for this). The electrical and atmospheric discharge projects must be done, and the installations must be reconstructed in an

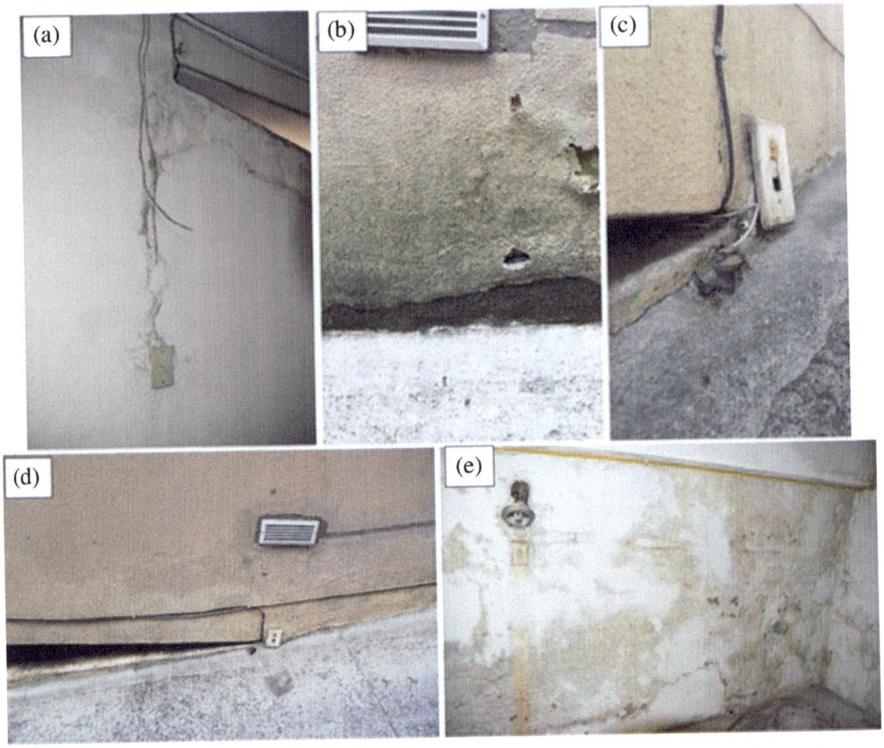

Fig. 28 Status of electrical and telephony installations, and antennas: **a** loose wires detached from the wall; **b**, **c** and **d** interventions to renovate the electrical installations and electrical gate of the garage; **e** poorly fixed lamp and presence of humidity and mustiness

efficient and safe manner [33, 34]. Rules must be structured in the condominium conventions and on user's manual with care instructions and clear rules for the services of these facilities.

5.4.2 Problems in Pluvial and Hydro-Sanitary Installations

Figure 29a–f show a lot of problems in pluvial, hydraulic and sanitary installations.

Prognosis: The hydraulic installations of the plumbs and reservoirs exit are old. In the apartments these installations were rebuilt, according to building's manager, however not yet in the common area. It is noticed the use of corroded metal pipes and connections with several leaks and in disagreement with cold water installations legislation [35]. The sanitary installations were in similar condition, due to several intervention works to unclog, and lack of preventive maintenance [36]. It highlights that part of the installation were concrete shackles. The pluvial issues (roof, slabs, ramps and open spaces) were worse due to several cracks in the floor which allow

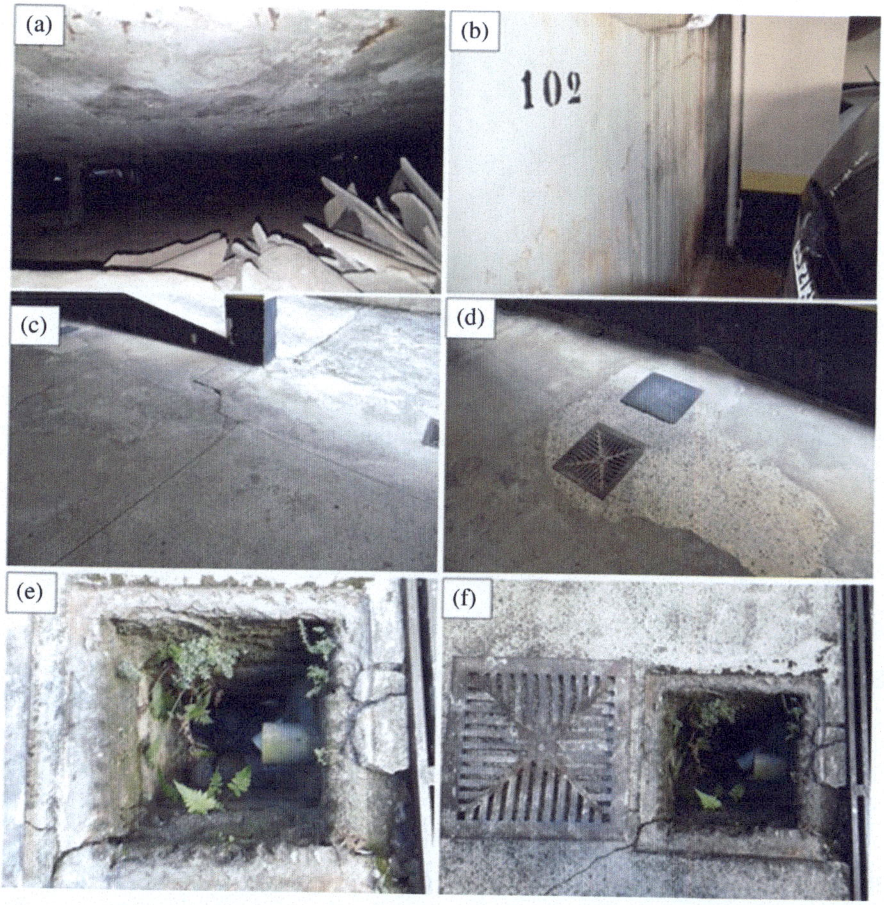

Fig. 29 Situation of hydraulic-sanitary systems: **a** leakage of the metallic installations of the plumbs of the upper reservoir exit; **b** rainwater stains from the roof of the garage; **c** several cracks on the floor where the rainwater passes through; **d**, **e** and **f** passage boxes and drains made of concrete with precarious waterproofing

water percolation, leveraging any damaged of the floor covering. In addition to these, there is the lack of waterproofing of drains and passage boxes of water and sewage, which are promoting water percolation and even vegetation proliferation, as seen in Fig. 29e.

Recovery Project: It is necessary to hire a qualified professional to assess the conditions of hydraulic plumbs and reservoir exit and rebuild them properly [35]. The sewage must be assessed and the parts with problems redone [36]. Rainwater collection (roof, slabs, ramps and uncovered areas) must be redesigned and remade in order to avoid it falling into the parking spots in the garage, and the passage boxes and drains must be properly waterproofed [20, 21]. The system with concrete shackles

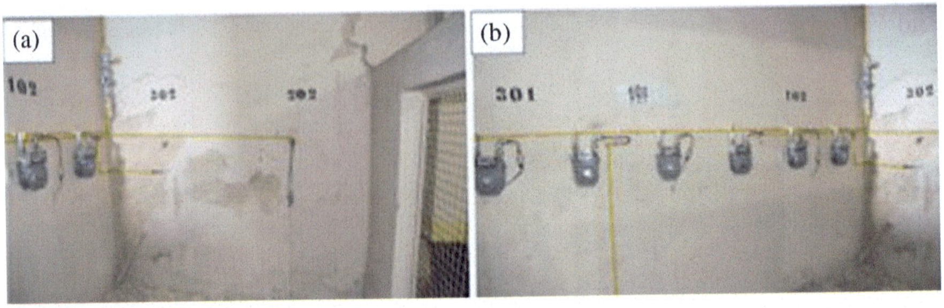

Fig. 30 Location of gas measurer: **a** side view; **b** front view

must be remade and changed for PVC. It is suggested to dry the lower reservoir and fill it in order to reduce empty spaces, accumulation of water and leakage into the soil.

5.4.3 Gas Installations Made Without Planning

Figure 30a, b show the location of the gas measurers.

Prognosis: Gas facilities already existed in the building (gas cylinders placed in the garage). The residents decided to change it for a pipe installation with individual measurers. Gas plumbing is provided on the street of the condominium. The installation was carried out by a gas supply company and it followed the recommendations of that company. The pipes and measurers were installed on a wall with humidity and powdery coating, which impairs its fixation.

Recovery Project: It is necessary to hire a qualified professional to assess the condition of the gas installations and their fixation after the reform of the previous items, as coating mortar. It must be ensured that facilities meet the standards [37].

6 Result of General Inspection and Priority Setting

All elements are evaluated in relation to the nine criteria, being classified as: lower (L), regular (R) or superior (S), as seen in Table 9. The method consists of scoring non-conformities to obtain the priorities. Initially, weights were assigned for each non-conformity in three functions: gravity, urgency, and tendency (see Table 7). The punctuation is obtained through the multiplication of these three weights. The priority is given in descending order, in other words the first event will be the event that has obtained the highest score, according to Table 10.

The method allowed to find the 10 most critical points which should be attacked first by the condominium:

Table 9 Inspection topics

Inspected areas		Technical conditions		Maintenance conditions		Use conditions			Administrative and financial management	
		Projects/documents	Functionality	Preventive actions	Corrective actions	Conservation and existence of anomalies	Environmental adequation	User risk	Administrative liabilities	Insolvency
		Projects/Manual	Sustainability	Preventive	Corrective	Integrity	Performance	Security	Administrative	Financial
Building	Structure	L	R	L	R	L	R	R	[a]	[a]
	Masonry	L	R	L	L	R	R	R	[a]	[a]
	Coating	L	L	L	L	L	L	L	[a]	[a]
	Floor	L	L	L	R	R	L	R	[a]	[a]
	Ceiling	L	R	L	R	R	R	R	[a]	[a]
	Covering/roofing	L	L	L	R	R	L	L	[a]	[a]
	Painting	L	R	L	R	L	L	R	[a]	[a]
	Electrical installations	L	L	L	R	R	L	R	[a]	[a]
	Hydro-sanitary installations	L	L	L	L	L	L	R	[a]	[a]
	Pluvial Installations	L	L	L	L	L	L	R	[a]	[a]
	Pavement	L	L	L	R	L	S	L	[a]	[a]
	Accessibility	L	R	L	L	R	R	R	[a]	[a]
Administration	Protocol register	[a]	[a]	[a]	[a]	[a]	[a]	[a]	[a]	[a]
	Assignee manual	[a]	[a]	[a]	[a]	[a]	[a]	[a]	[a]	[a]

(continued)

Table 9 (continued)

Inspected areas		Technical conditions		Maintenance conditions		Use conditions			Administrative and financial management	
		Projects/documents	Functionality	Preventive actions	Corrective actions	Conservation and existence of anomalies	Environmental adequation	User risk	Administrative liabilities	Insolvency
		Projects/Manual	Sustainability	Preventive	Corrective	Integrity	Performance	Security	Administrative	Financial
	Condominium meetings	[a]	[a]	[a]	[a]	[a]	[a]	[a]	[a]	[a]
	Invoices	[a]	[a]	[a]	[a]	[a]	[a]	[a]	[a]	[a]
	Labor documents	[a]	[a]	[a]	[a]	[a]	[a]	[a]	[a]	[a]
	Finance register	[a]	[a]	[a]	[a]	[a]	[a]	[a]	[a]	[a]

[a]It was not possible to evaluate, due to the lack of documents or information about the subject

Table 10 Worksheet of priorities

Unconformities		Severity	Urgency	Tendency	Score	Priority
5.1	*Infiltrations and pathological manifestations by humidity*					
5.1.1	Lack of sealing of the building's roof and the upper water tank	6	10	10	**600**	7°
5.1.2	Lack of waterproofing in the roofs of the garage and entrance hall	6	10	10	**600**	8°
5.1.3	Moisture from splashing on the floor	3	10	3	**90**	18°
5.1.4	Moisture from inefficient waterproofing	6	10	8	**480**	11°
5.2	*Pathological manifestations in the structural elements of the building*					
5.2.1	Horizontal cracks in the garage's columns	10	10	8	**800**	3°
5.2.2	Cracks close to the longitudinal reinforcement and concrete swelling	10	10	10	**1000**	1°
5.2.3	Cracks in the encounter of structures with different ages	10	10	10	**1000**	2°
5.2.4	Inefficient stair handrail	10	10	8	**800**	4°
5.3	*Pathological manifestations on building coatings*					
5.3.1	Dirt stains on coatings	3	10	10	**300**	14°
5.3.2	Detachment and retraction of coatings	6	10	10	**600**	9°
5.3.3	Disaggregation and powder behavior of coatings	6	10	10	**600**	10°
5.3.4	Vegetation fixed on the building wall	3	10	6	**180**	17°
5.3.5	Lack of coating on boundary wall at the back of the building	6	10	6	**360**	12°
5.3.6	Inappropriate coat for columns and beams of the garage	3	10	6	**180**	15°

(continued)

Table 10 (continued)

Unconformities		Severity	Urgency	Tendency	Score	Priority
5.3.7	Aged painting system	3	10	6	**180**	16°
5.3.8	Lack of lintels and counter-lintels in doors and windows	3	10	3	**90**	19°
5.3.9	Damaged ceramic floors in the building facilities	3	10	3	**90**	20°
5.4 Pathological manifestations in the building facilities						
5.4.1	Problems in electrical, telephony and antennas installations	8	10	8	**640**	5°
5.4.2	Problems in pluvial and hydro-sanitary installations	8	10	8	**640**	6°
5.4.3	Gas installations made without planning	6	10	6	**360**	13°

1°) 5.2.2—Cracks close to the longitudinal reinforcement and concrete swelling;
2°) 5.2.3—Cracks in the encounter of structures with different ages;
3°) 5.2.1—Horizontal cracks in the garage's columns;
4°) 5.2.4—Inefficient stair handrail;
5°) 5.4.1—Problems in electrical, telephony and antennas installations;
6°) 5.4.2—Problems in pluvial and hydro-sanitary installations;
7°) 5.1.1—Lack of sealing of the building's roof and the upper water tank;
8°) 5.1.2—Lack of waterproofing in the roofs of the garage and entrance hall;
9°) 5.3.2—Detachment and retraction of coatings;
10°) 5.3.3—Disaggregation and powder behavior of coatings.

This is coherent, as they represent the most critical part of the building and which require hiring a structural professional for better assessment and projects development, and the construction of structural reinforcement, urgently.

The other part that was identified as more serious, is related to the conditions of the roof and building's waterproofing system that also need to be seen urgently, to avoid it impacting on structural system due to humidity presence.

It is important to highlight the conditions of installations that pose a risk of electrical shock and breakdowns due to absence of projects and user's manuals. As well as the drinking water that may be contaminated due to pipes corrosion. It is hoped that this chapter allow readers to better structure their maintenance and recuperation schedules for all building's structures.

References

1. A.C. Lordsleem Jr., H.F. Faro, Building façade cladding detachment: a case study. Revista ALCONPAT. **7**(2), 148–159 (2017)
2. C. Pereira, E. Silva, A. Hamadyk, Probabilistic analysis of the durability of architectural concrete surfaces. Appl. Mathem. Mod. **77**, 199–215 (2020)
3. C. Bocchile, Labor safety: capital against the risk. Construção Mercado. **9**, 29–34 (2002). (in Portuguese)
4. A.T. Souza, L.A. Riccio, G.C. Laquini, W.J. Santos, Behavior of mortar coating subjected to extreme conditions: lack of curing and no substrate moistening. Int. J. Sci. Eng. Investig. **7**, 53–59 (2018)
5. W.J. Santos, E.V.M. Carrasco, J.N.R. Mantilla, E.M. Piancastelli, A.G. Magalhães, F.J. Silva, M.A.P. Rezende, Study of the effects of excessive cement or lime in mechanical properties and durability of the mortar coating. Int. J. Dev. Research. **09**, 29923–29927 (2019)
6. J.S. Souza, E. Bauer, M.L.M. Nascimento, V.M.S. Capuzzo, V.A.G. Zanoni, Study of damage distribution and intensity in regions of the façade. J. Build. Pathol. Rehabil. **3**, 1–9 (2016)
7. C. Pereira, A. Silva, J. Brito, J.D. Silvestre, Urgency of repair of building elements: prediction and influencing factors in façade renders. Constr. Build. Mater. **249**, 1–16 (2020)
8. WHO Regional Office for Europe, *Guidelines for Indoor Air Quality: Dampness and Mould* (World Health Organization, Denmark, 2009)
9. P.S. Kian, A Review of factors affecting building defects in Singapore. Dimens. Tek. Sipil. **3**(2), 64–68 (2001)
10. M.Y.L. Chew, Defect analysis in wet areas of buildings. Constr. Build. Mater. **19**, 165–173 (2005)
11. U.S. Environmental Protection Agency, Moisture Control Guidance for Building Design, Construction and Maintenance. EPA, Washington, DC (2013)
12. A.J. Alma, K.R. Lis, H.O. Hygen, An approach to impact assessments of buildings in a changing climate. Build. Res. Information. **39**, 227–238 (2011)
13. M.G.H. Lima, Diagnosis of incidences of pathological manifestations using the gut matrix methodology—case study. Monograph (Federal Rural University of the Semi-Arid) of Civil Engineer (2019)
14. H. Rozenfeld, F. Forcellini, D.C. Amaral et al., *Product Development Management: A Reference for Process Improvement* (Saraiva, São Paulo, 2006)
15. M.S. Ribeiro, C. Michalka Jr., The contribution of industrial construction processes to the adoption of new technologies in civil construction in Brazil. Revista Vértices. **5**(3), 89–108 (2003). (in Portuguese)
16. Brazilian Institute of Evaluations and Engineering Expertise—IBAPE, National building inspection standard. IBAPE, Sao Paulo (2012) (in Portuguese)
17. Brazilian Association of Technical Standards—ABNT, NBR 5674: Building maintenance—Requirements for maintenance management system. ABNT, Rio de Janeiro (2012)
18. Brazilian Association of Technical Standards—ABNT NBR 14037: Guidelines for building use, operation and maintenance manual preparation—Requirements for content preparation and presentation. ABNT, Rio de Janeiro (2014)
19. Brazilian Association of Technical Standards—ABNT, NBR 16373: Steel roofing and thermoacoustic panels—Performance requirements. ABNT, Rio de Janeiro (2015)
20. Brazilian Association of Technical Standards—ABNT, NBR 9575: Waterproofing—Selection and project. ABNT, Rio de Janeiro (2010)
21. Brazilian Association of Technical Standards—ABNT, NBR 9574: Execution of Waterproofing. ABNT, Rio de Janeiro (2008)
22. Brazil, Regulatory standard 18—Health and safety conditions at work in the construction industry. Ministry of Economy, Brasilia (2020)
23. Brazil, Regulatory standard 35—Work Safety at Height. Ministry of Economy, Brasilia (2020)
24. Brazilian Association of Technical Standards—ABNT, NBR 5626: Cold water building installation. ABNT, Rio de Janeiro (1998)

25. Brazilian Association of Technical Standards—ABNT, NBR 7200: Render made of inorganic mortars applied on walls and ceilings—Procedure for application. ABNT, Rio de Janeiro (2013)
26. Brazilian Association of Technical Standards—ABNT, NBR 13245: Paints for buildings—Painting of non-industrial buildings: surface preparation. ABNT, Rio de Janeiro (2011)
27. Brazilian Association of Technical Standards—ABNT, NBR 6118: Design of concrete structures—Procedure. ABNT, Rio de Janeiro (2014)
28. Brazilian Association of Technical Standards—ABNT, NBR 14718: Frames—Bodyguards for edification—Requirements, procedures and test methods. ABNT, Rio de Janeiro (2019)
29. Brazilian Association of Technical Standards—ABNT, NBR 15575: Residential buildings. ABNT, Rio de Janeiro (2013)
30. Brazilian Association of Technical Standards—ABNT. NBR 10821: Frame for buildings. ABNT, Rio de Janeiro (2017)
31. Brazilian Association of Technical Standards—ABNT. NBR 13529: Inorganic mortar coating for walls and ceilings—Terminology. ABNT, Rio de Janeiro (2013)
32. Brazilian Association of Technical Standards—ABNT (2001) NBR 9077: Buildings—Emergency exits—Procedure. Cold water building installation. ABNT, Rio de Janeiro (2013)
33. Brazilian Association of Technical Standards—ABNT. NBR 5410: Electrical installations of buildings—Low voltage. ABNT, Rio de Janeiro (2004)
34. Brazilian Association of Technical Standards—ABNT. NBR 5419: Lightning protection. ABNT, Rio de Janeiro (2018)
35. Brazilian Association of Technical Standards—ABNT. NBR 5656: Cold water building installation. ABNT, Rio de Janeiro (1998)
36. Brazilian Association of Technical Standards—ABNT. NBR 8160: Sewage building systems—Design and installation. ABNT, Rio de Janeiro (1999)
37. Brazilian Association of Technical Standards—ABNT, NBR 15526: Internal network distribution for fuel gases in residential installations—Project and execution. ABNT, Rio de Janeiro (2016)

Comparative Study of Consumption and Life-Cycle Impacts of Water Heating Systems for Residential Multi-familiar Buildings in Rio de Janeiro, Brazil

Arthur B. Silva, Mohammad K. Najjar, Ahmed W. A. Hammad, Assed Haddad, and Elaine Vazquez

Abstract Civil construction, as an integral part of the chain of industrial activities, is one of the various segments responsible for energy consumption and potential greenhouse gas emissions throughout its life cycle. The building materials and their systems have a direct influence on energy consumption and impact assessment. In this context, Hot Water Building Systems (HWBS) are included. The type of installation to be used in a building is defined by technical and/or economic requirements. However, the spectrum of possibilities should consider resource consumption and generation of environmental impacts throughout the life cycle. This research proposes a novel application of an environmental management method to empower the decision-making process of HWBS, insights a Life Cycle Assessment (LCA) methodology to compare the environmental performance of two distinct HWBS (i.e. Natural Gas Heating System and Solar Heating System) for multi-family residential developments. In conclusion to the results obtained, it can be inferred that the HWBS with heating via SHS has better environmental performance than the system with heating via Natural Gas, even though the first one uses an electrical complement for operation appropriate.

Keywords Life cycle assessment · Water heating systems · Residential buildings

A. B. Silva (✉) · A. Haddad
Programa de Engenharia Ambiental, Universidade Federal do Rio de Janeiro, 21941-909 Rio de Janeiro, Brazil
e-mail: bastos.arthur@poli.ufrj.br

A. Haddad
e-mail: assed@poli.ufrj.br

M. K. Najjar
Centro Universitário Gama e Souza (UNIGAMA), 22621-090 Rio de Janeiro, Brazil
e-mail: mnajjar@poli.ufrj.br

A. W. A. Hammad
Faculty of Built Environment, University of New South Wales, 2052 Sydney, Australia

E. Vazquez
Programa de Engenharia Urbana, Universidade Federal do Rio de Janeiro, 21941-909 Rio de Janeiro, Brazil
e-mail: elaine@poli.ufrj.br

J. M. P. Q. Delgado (ed.), *Case Studies in Building Constructions*, Building Pathology and Rehabilitation 15, https://doi.org/10.1007/978-3-030-55893-2_7

1 Introduction

Population growth and technological development in recent decades have made human activities largely responsible for structural changes in the environmental landscape at regional and global levels [1, 2]. Regarding the aspects of natural resources consumption and the passive impact of human activities, the energy sector is responsible for a major part of greenhouse gas (GHG) emissions [3]. For instance, residential and commercial buildings account for around 41% of total energy consumption in the United States [4]. In these terms, building components have a direct influence on energy consumption and environmental impacts over the entire Life Cycle Assessment (LCA), basically, during the pre-operational phase (i.e. material manufacturing, transportation and construction), as well as at the end-of-life and disposal phase [5].

Buildings are major consumers of energy throughout their life cycle. Generation of energy primarily depends on conventional sources, which is the basic cause of environmental pollution [6]. The materials and their systems have a direct influence on energy consumption and impact generation, in the pre-operational phase (materials manufacturing, transportation and construction), also in the final and discarded life.

Hot Water Building Systems (HWBS) are directly related to energy consumption in residential buildings; performing the second largest energy consumer in buildings and, thus, representing an integral part of the water-energy nexus [3].

The conventional selection of a water heating system in residential buildings focuses on the financial evaluation rather than the sustainability pillars and life cycle consequences (i.e. economic × environmental pillars) [7]. At this level of the analysis, the application of LCA methodology at an early designing phase of residential buildings could empower the decision-making process and sustainability [8], as well as facilitating the selection criteria of HWBS [9], where professional and experts could evaluate the environmental performance of the installed water heating system [10, 11].

A novel application of an environmental management method is presented herein to empower the decision-making process and encourage the selection course of HWBS, taking into consideration the technical and economic aspects at an early designing phase of buildings. The aim of this work is to present a proposal for a method derived from the general LCA methodology in order to compare the environmental performance of two distinct HWBS for multi-family residential developments, through thermal heaters installed on the final roof of buildings, with supplementation of electrical supply, so that accurate information on the environmental performance of the systems can be obtained. However, the installed HWBS considered in this work are NGHS and SHS. In this work, a literature review of the LCA methodology is presented in Sect. 2. The proposed methodology to evaluate the LCA for HWBS is presented in Sect. 3. However, results are discussed in Sect. 4, while the conclusions and final recommendations are presented in the last section.

2 A Literature Review of LCA Methodology

LCA is described by as a scientifically based analysis and assessment of the environmental impacts of product systems [12]. Regarding to ISO standards, this methodology was revised in 2006 and started to be condensed into ISO 14040 and ISO 14044 standards [13]. In Brazil, the Brazilian Association of Technical Standards, published equivalent versions initially translated in 2001 (NBR ISO 14040, NBR ISO 14041, NBR ISO 14042 and NBR ISO 14043) and later in 2009 and 2014 (NBR ISO 14040 and NBR ISO 14044 replacing the previous ones) in a way to support the descriptive text and definitions, as well as facilitate the understanding of the theme [14].

At the level of the energy consumption and impact assessment of products, LCA is characterized as a management methodology that help computing inputs and outputs of a production system to evaluate the environmental performance over their entire lifespan [15]. The application of LCA methodology in the construction sector focuses mainly on the characteristics of the building typology and components [16]. However, such application is facing several challenges that are giving a wide spectrum of related variables and making it necessary and interesting to define a standardized analysis structure in order to increase its accuracy [17]. In this context, HWBS should also be assessed over their entire lifespan, so that the energy incorporated into the biogenic emissions are considered and give greater dimension to the impacts of the systems [18]. The methodology adopted for this study is related to the LCA of HWBS in multifamily residential buildings, taking into consideration comparing the different types of systems in relation to their environmental performance at an early designing step. The general scope of such application will be conducted in four phases based on the LCA methodology: Goal and Scope definition; Life Cycle Inventory analysis (LCI); Life Cycle Impact Assessment (LCIA); and Interpretation of data and results obtained by the partial and final methodological processes [19, 20].

Defining the Goal and Scope of the study means determining the intended application for the analysis and the reason for carrying out the study, the target audience to whom the results are intended to be communicated and, therefore, the means for their dissemination [19]. In these terms, the scope of the study should include the definition of the product system to be studied; the functions of this system or compared systems; the determination of the functional unit/functional equivalent; the system boundary; allocation procedures; selected impact categories and methodology for impact assessment as well as subsequent interpretation to be used; data requirements; assumption; limitations; initial requirements for data analysis; type of critical analysis, if applicable; type and format of report required for study.

The next step is to build up the LCI based on ISO 14044, which demonstrates a definition of the inventory analysis phase that involves the cradle-to-grave character of the method; "life cycle assessment phase involving the compilation and quantification of inputs and outputs of a product system over its lifetime. life cycle". Table 1 illustrates the output results of the LCI step [21], including a list of data of the environmental impacts to be evaluated at the next step of the LCA methodology,

Table 1 Characterization of impact categories commonly demonstrated in studies

Impact categories	Geographic scale		
	Global	Regional	Local
Global warming	×		
Stratospheric ozone depletion	×		
Phoyochemical oxidant formation		×	×
Acidification		×	×
Nutrient enrichment		×	×
Ecotoxicity		×	×
Human toxicty		×	×
Working environment			×
Odour			×
Noise			×
Radiation	×	×	×
Resource consumption	×	×	×
Land use			×
Waste			×

Adapted from Stranddorf and Hoffman [21]

which is LCIA. At this level of the analysis, LCIA aims to give an overview of the significance of the potential impacts of the examined product [22].

There are several methods to evaluate the extracted impacts from the LCI, hence, it is highly important to choose the most appropriate method for each case study [18]. LCIA can be distinguished within two levels: midpoints and endpoints [23]. At the midpoint impact assessment level, indicators are given along the environmental mechanism, while at the endpoint impact assessment level, "*Characterization considers the entire mechanism to its end point, i.e. it refers to a specific damage related to the broader area of protection, which may be human health, natural environment or natural resources*" [24]. Finally, the interpretation level refers to permeate the entire analysis process, where the findings of LCI and LCIA are to be consistently combined with the defined Goal and Scope in order to draw conclusions and recommendations [19].

3 Proposed LCA Methodology for HWBS

The objective of this work is to Present a methodological work flowchart for the comparative application of LCA for HWBS in multi-family residential buildings, as a way to obtain data for analysis regarding the environmental impacts of such systems, taking into consideration the energy consumption during of the operation phase of HWBS and its impact on the living standards, combined with the LCA database of the of the applied materials. The developed method herein is based on nine main phases, which guide the elaboration and evaluation of the projects and their respective analysis and are the organization of the general phases of methodological application of the LCA recommended by ISO Standards. At this level of the analysis, the interpretation phase is divided into four distinct stages; Interpretation (A), Interpretation (B), Interpretation (C) and Interpretation (D), which means that in each phase, I, II, III and IV in the end of it will have an interpretation of the results. Each stage has been oriented and modified to verify the output results collected from the previous steps (LCI and LCIA), as well as evaluating their quality, coherence and importance to the study. Figure 1 describes in detail the steps contained in phases I, II, III, and IV of the study, which are Goal and Scope Definition (A), Interpretation (A), Goal and Scope Definition (B), and Interpretation (B), respectively.

The starting point of the analysis, as presented in Fig. 1 is to determine the environmental profile of a certain HWBS. Phase I, Goal and Scope Definition (A), means to conduct general definitions, function, and functional unit. At this level of the analysis, Item (I.1), presented in Fig. 1, means determining the general purpose of the analysis to be performed. Such an objective must be clear and consistent with the reality of the place of application so that it can be valid and have real importance in the context in which it will be applied. The results of the analysis of the performance of the HWBS during the operation phase can be used to make a decision about the use of a particular type of system still in the design phase, with the general objective being traced, for example, as the definition of the type of system, system to be used for the distribution of water in a specific building type, or if a copper or PVC pipe is defined in the project, given the observed impacts.

Item (I.2), presented in Fig. 1, means determining the target audience. In the case of the building installation project, the target audience can be defined as the end user, who will actually use the system and wants to know which one is most advantageous in this respect, or the builder who will do the work and needs the best cost-benefit, or it may be the own design team that needs the determination of the system that consumes the least environmental resources or generates the least impact in order to have a sustainable building profile that seeks environmental certification. Next, item (I.3) determines the scope of the analysis. At this point, a phase of refinement of objectives is to be included, defining the stages of application, scope, work team involved and other important aspects to the elaboration. Moreover, the evaluated function in the employed method can be traced in item (I.4). The main functions of cold water, hot water and piped gas systems can be analyzed, considering that the functions are closely linked to the uses of systems. The functional unit of the study, presented in item (I.5) in Fig. 1, deals with the quantification of the determined function or product performance characteristic. In these terms, the functional unit must, in the case of building systems, be worked as a performance unit and not a mass or metric unit.

Phase II, Interpretation (A), refers to identify stakeholders, focusing on a more managerial profile of the process. Construction projects involves many stakeholders (i.e. facility owners and users, project managers, project team members, facility managers, designers and architects, allotment companies, shareholders of

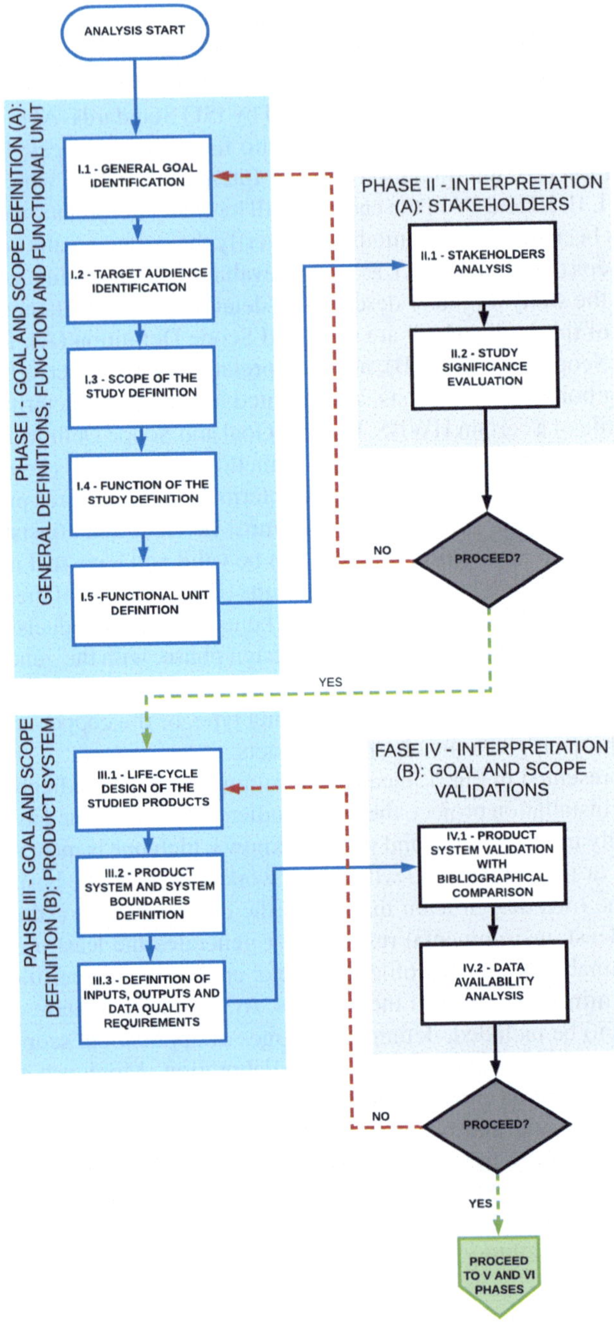

Fig. 1 Detailed flowchart of the proposed methodology of this work—Phases I to IV

the company that develops the enterprise, the public administration, construction workers, subcontractors and outsourced service providers, competitors, banks, insurance companies, representatives of the surrounding community, the general public, and others) [25].

Phase III, Goal and Scope Definition (B) begins by defining the product system, inputs and outputs, which determines the life cycle design of the studied facility systems, as well as the system boundaries. At this level of the analysis, it is necessary to trace the processes and phases involved in the life cycle of the hydraulic installations from the production of the component materials to the final disposal of the system after the operation phase. Conducting the life cycle design of the studied products leads to item (III.2), where the product system and system boundary are to be determined. Item (III.3), presented in Fig. 1, necessitates defining the inputs and outputs of the product system and data requirements. In this work, the most important issue when building up the inventory of database is the quality, relevance, accuracy completeness, and representativeness of the data due to technological and profile changes of the products used to HWBS, taking into account the type and location of installation, consistency and reproducibility of the products.

Phase IV, Interpretation (B), refers to validate the product system. For this, a comparison should be made between the determined life cycle and product system for analysis with other similar related studies and make adjustments that the professional deems necessary to make the system as objective as possible with respect to the results. This factor is extremely important to the legitimacy of the study, since the geographical location and the local social and environmental profile have a direct influence on the quantification and qualification of the impacts of a product system represented by a building installation.

The second sequence of the detailed flowchart of the proposed methodology if this work is illustrated in Fig. 2, represented by Preliminary System Design Development (Phase V), Inventory Analysis (Phase VI), and Interpretation (C) (Phase VII).

Phases V and VI should begin concurrently or sequentially, as the first step in the inventory phase is to define and organize data sources to enable collection, organized according to requirements. In order to obtain these, it is oriented to use an internationally consolidated database, since the exact obtaining of the production processes involved denotes time and resources. However, such a process may lead to variability of results with respect to the actual life cycle of the examined system and compromise the reliability and requirements of the data. Data collection at this level of the analysis may involve the need for bibliographic and market research that fosters the assembly of product system processes and guides the volume of materials consumed for them.

Phase V consists of the preliminary design of building systems for analysis objects for the primary purpose of obtaining the data necessary to foster life cycle inventory analysis, as presented in Fig. 2. Hence, four main sequential activities related to good engineering practices are to be defined, as follows:

An architectural assessment of the needs and demands of the building should be carried out. The project layout should be defined, taking into consideration elements

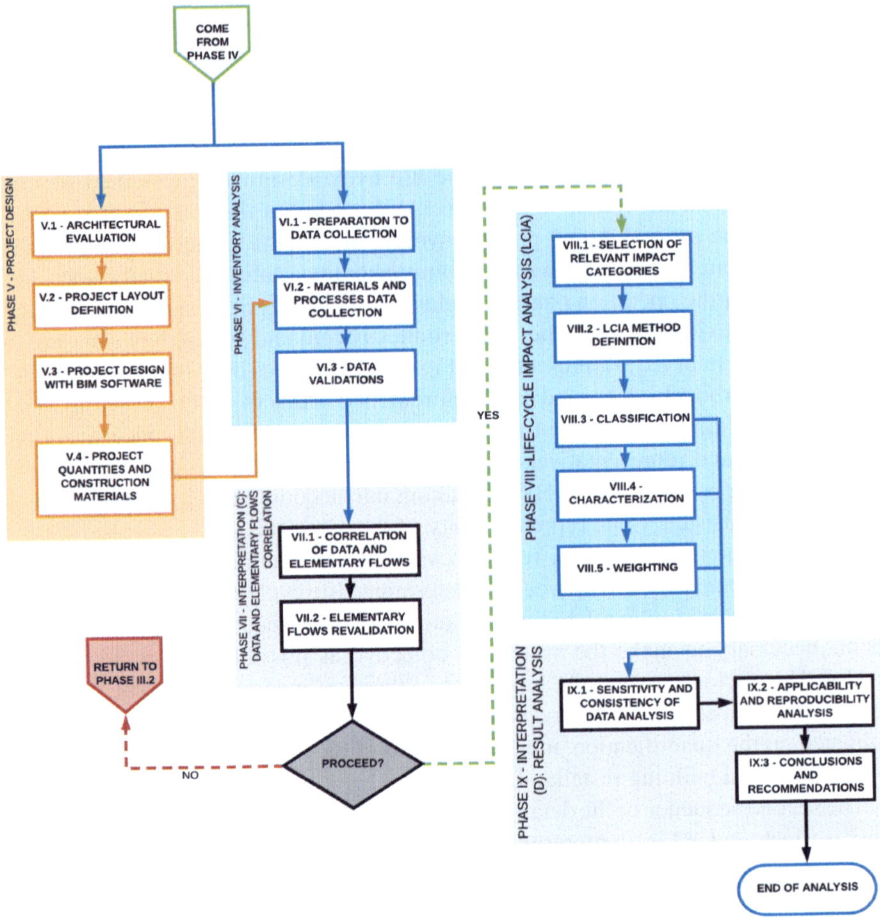

Fig. 2 Detailed flowchart of the proposed methodology of this work—Phases V to IX

such as technical reserve location, water or power inlet leasing, internal distribution pipe division, better leasing of equipment needed for system operation.

The effective elaboration of the projects should be considered based on the current norms and available database. Hence, one should seek to understand the level of details that the project necessitates to meet the requirements of the LCA methodology such as the specifications of buildings components, taking into consideration that the LCA methodology involves processes such as "cradle to grave", that is, from the extraction and manufacturing phase until the end of life; "cradle to gate", that is, from the extraction until the end of manufacturing phase; "cradle to cradle", that is, from the extraction and manufacturing until the end of life and recycling to be reused again [13]. The project materials and construction components should be quantified, considering the use of the products that will actually be installed to increase the reliability and completeness of the data obtained.

Phase VI, inventory analysis, is the elaboration phase of the projects that ends up with obtaining the list of materials that fosters the inventory data. The data collection requires obtaining data related to the life cycle of the components of the HWBS; a factor that is intended to be performed with the help of a database, such as Ecoinvent. This phase necessitates guiding the modeling of the product system with the aid of LCA software, such as OpenLCA. System modeling often involves the combination of basic processes and raw materials in the database to obtain the desired products [26], a factor that can create uncertainties in the process, given the insufficient knowledge obtained about the process. Production or misuse of processes. At this level of the analysis, the data collection stages consider the database collections, materials and processes data collection, and culminate in the data validation. The data collected in the inventory should be evaluated against the data requirements defined at the beginning of the study in a way to determine the relevance or discard of collected data by screening the material.

Phase VII is Interpretation (C), where the correlation of data and elementary flows is ranked in the classification of the collected data according to the defined flows for the studied functional unit [27]. Thus, it is verified whether all flows considered have consistent and sufficient data for the elaboration of the LCIA, if other data are needed or if there is not enough data available for the definition of all flows. At the end of this analysis, a decision should be taken whether to continue the process or to redefine and redo the completed phases to ensure concise results. The carried inventory and the refined product system facilitate proceeding to the LCIA and final interpretations, which are detailed in the figure, Fig. 2, which demonstrates the final phases of the proposed methodology of this work.

Phase VIII, LCIA, means selecting the relevant impact categories to the study, taking into consideration the history of application of LCA for analysis of hydraulic systems. The major impact categories, as previously presented in Table 1, can be exposed to global warming, human toxicity (carcinogenic and non-carcinogenic), shortage of fossil resources and mineral resources and waste, impacts considered directly related to the building systems of employment of the study. The next step after selecting the impact categories, the LCIA step is to be conducted according to the item (VIII.2), illustrated in Fig. 2. Hence, it is important to consider the midpoint impact assessment, which has less data uncertainties [28], using an impact assessment method such as ReCiPe, which combines the Eco-indicator 99 and CML methods, giving them an update regarding the content, deriving characterization factors according to a midpoint approach (with 17 indicators) or endpoint (with 3 indicators) [29]. There is a fundamental relationship between midpoint indicators, direct impacts, and endpoint indicators. The structure overview of ReCiPe impact assessment method is presented in Fig. 3 [30]. Phase IX, Interpretation (D), which necessitates evaluating the completeness, sensitivity and consistency of the data.

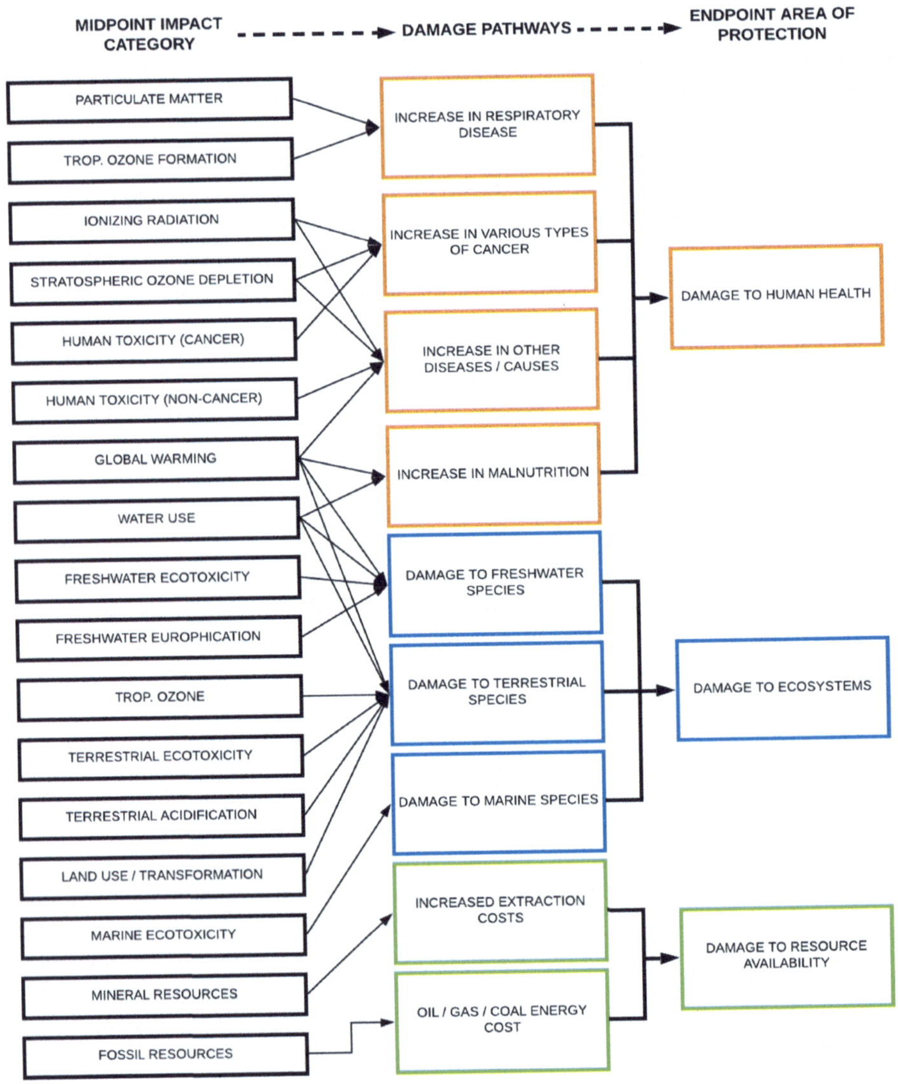

Fig. 3 Structure overview of ReCiPe impact assessment method (Adapted by RIVM [30])

4 Results

A multifamily residential project designed with five buildings interconnected by patio and access corridors, called "Quilombo da Gamboa", was chosen as a case study, which will constitute a housing complex with 116 apartment units.

The main definitions of step (I) are presented, with the objective, scope and functional unit. The general objective was to apply the life cycle assessment method to obtain environmental performance information for a hot water building system with natural gas heating, compared to a hot water building system with a thermal solar heating source, taking into account the executive particularities of the project and the peculiarities of the site. Already the scope of work in relation to the application of LCA methodology for two situations of HWBS's arrangements, considered the relevant equipment installed, the layout and preliminary sizing of pipes, as well as launching the positions of the consumer devices (considering only the showers of bath in each apartment), appliances and heating systems. On the other side, consumption of hot water for cooking, washbasins and other secondary activities was neglected, taking into consideration that it is desired to obtain only the heating performance related to the bath, which is the biggest consumer of water and electricity converted into thermal in homes in the country. The pre-operational and operational phases of the chosen building systems were considered, defining a useful life of 25 years. Finally, the functional unit for the study was configured as a performance unit, fixed according to the parameters necessary for the analysis. In line with the above, the functional unit was defined as the "volume of water at a temperature of 40 °C, necessary for a bath with an average duration of 5 min, in a shower installed in the system, for an estimated lifetime of 25 years". At this level of the analysis, phase (II) was not carried out, taking into consideration that the presented projects were not submitted to an effective decision making by which party should be taken in the case of the hot water installations of the buildings.

The product systems in the pre-operational and manufacturing stage and an operational stage were defined in phase (III). After defining the product systems, the study proceeded with the step of determining the reference flows (building materials of the system and the processes and energy related to its manufacture and transportation; energy and water consumed; environmental interventions), related to the inputs and outputs of the processes involved in the proposed systems. In phase IV, for the validation of the system, it is concluded that for the application of the production processes of the building system component materials, the Brazilian market, there were no databases that sufficiently cover the entire chain. Thus, this work opted for the use of the global market, within Ecoinvent, widely used for LCA studies by the academic community.

In phase V, the projects for the two study systems were elaborated; the installation of the Hot Water System by Gas Heating and via Solar Thermal Heating. For the first system, the design of the bath water heating system was structured so that the supply starts with the connection of the system with the public network of the local concessionaire, installed in the public logotype, and the from the general supply, individualized measurement of gas consumption for heating of each residential unit belonging to the set of buildings is carried out. The gas heating system for the housing complex was made up of individual passage heaters for each unit. For the second system, the project was structured from a solar collection system, with solar collectors, installed in the highest location of the building with water supply for heating through an exclusive branch derived from the upper reservoir of cold water available for each building block. The water will be destined to the collection of solar collectors through copper tubing with thermal insulation, and will be forced to circulate, with a centrifugal circulation pump. For the hot water reserve, a general thermal reservoir located on the roof will be provided, from which the apartments' supply barrel will be derived, which will have columns of exclusive CPVC pipe[1] hot water, with consumption meters. The system was completed with the use of an electric shower.

Phase VI represents the analysis of the inventory in line with the developed project, gathering information on the building materials of the systems, considering a pre-operational stage, which goes back to the extraction of raw materials, manufacturing and transportation of the systems. Components up to the construction of the system and other related to the operation itself, which involves energy consumption during the life of the system, eventual maintenance and other factors. Hence, the OpenLCA software and the Ecoinvent database were used in the modeling of the product system.

For the HWBS inventory with heating via natural gas according to the definitions of the product system and its borders, the components considered were: CPVC piping for the conduction of the heated water from the heater to the point of consumption, copper tubing for conduction of natural gas, passage heating devices (aluminum and copper), shower installed in the bathrooms of the apartments (painted ABS), gas meter installed at the entrance of the branch (aluminum alloy) and closing record of the same (aluminum alloy) copper). In relation to the manufacturing processes associated with the materials, it was considered: for CPVC piping—PVC chlorination, extrusion and injection (for connections), for copper piping and passage heating devices—extrusion and drawing for showers—injection and chrome plating, for gas meters—pressure casting and for registers—casting and machining.

For the HWBS inventory with solar thermal heating and electrical complementation according to the product system definitions and its borders, the components considered were: CPVC pipe for water circulation for heating by solar collectors, PVC pipe for drinking water supply from the heating system, solar collectors, thermal reservoir with different capacities for reserving and reheating water with electrical supply and internal resistances, centrifugal pumps for forced circulation of water by the heating system, hydrometer for measuring the hot water consumption of apartment columns, column closing record and shower for use. Regarding the manufacturing processes associated with the materials, the following considerations have been done as follows: for CPVC piping—PVC chlorination, extrusion and injection (for connections), for PVC piping—PVC chlorination, extrusion, injection (for connections) and chrome plating, for solar collectors, thermal quenching, molding, extrusion, casting, centrifugation, polymerization and molding, extrusion and drawing, for reservoirs—lamination, assembly, casting and centrifugation, for pumps—casting, machining and assembly, for water meters—casting, for records—chlorination of PVC and injection and for showers—injection and chrome plating.

Phase VII was initially determined to define the absolute values applied to the functional unit, which considers only one shower installed in the system consisting

[1] Hot water piping systems in Brazil are usually built with CPVC (Chlorinated Polyvinyl Chloride) pipes, a polymer obtained by the industrial chlorination of PVC (Polyvinyl Chloride), which is used as material in the fabrication of pipes for cold water systems. The chlorination process gives the material high temperature resistance properties.

of 116 units. These data were considered as inputs of the product system modeled within the OpenLCA software. In phase VIII, regarding the calculation of impacts, the ReCiPe midpoint method was considered, which encompasses 18 different impact categories, of which the most important categories were defined for analysis: global warming, human toxicity (carcino-genic and non-carcinogenic), shortage of fossil resources and mineral and waste resources, impacts considered directly related to the building employment systems of the study. However, phase IX presented the midpoint impact categories compared between the two systems, in normalized values. At this level of the analysis, a normalized result for the ReCiPe midpoint impact categories is presented in Fig. 4.

It can be realized that for 14 categories presented in Fig. 4 that the impacts generated by the HWBS with gas heating are higher, for the considerations made in the study. HWBS with solar heating and electric complement only contributes more to the impacts of water consumption, land use, carcinogenic human toxicity and scarcity of fossil resources. The factor corroborates with expected results from the bibliography, insofar as the SHS consists of a source of clean main energy, without consuming fossil resources, for example. Even so, factors related to the production chain of component materials contributed strongly to several categories of impact in both systems. With regard to water consumption, it is noted from the analysis of the results of the category that the greatest contribution to the consumption of this resource by the SHS consists in the use of hydroelectric energy matrix electricity. The result demonstrates that the SHS tends to be less polluting than the gas source, however, in a practical application, in which there is a need to combine it with another energy source so that the system fully meets the demand, performance environmental impact on factors such as water consumption.

The impact category represented by human carcinogenic toxicity consists of another factor in which the SHS has greater influence, a factor associated with the treatment of residues from the production processes of the components, mainly the solar collector, such as red mud residue from bauxite production. With regard to the scarcity of fossil resources, the impacts generated by the HWBS with heating via NGHS represent only (11.49%) of the impacts generated by the system with SHS, a factor directly associated with the pipe used for the heating system via SHS, the HWBS, composed with polymer produced from the cracking of oil. Among other categories of relevance to the systems is global warming, measured by equivalent CO_2 emissions. It is noted that the carbon emission of the SHS represents approximately (82%) of the emission obtained with the heating system via NGHS.

Regarding the impact category represented by the scarcity of mineral resources, the impact analysis shows that the system with heating via NGHS contributes more than twice as much to the impact as the system via SHS, a factor directly associated with the use of the pipe copper for gas supply and distribution. It is shown that the greatest contribution to the impacts generated by the SHS consists of the life cycle of the collectors (60.72%), a factor directly linked to the pre-operational phase, and that the operational phase makes little contribution to the presented impact categories. On the other side, the system with heating via Natural Gas has a strong influence of

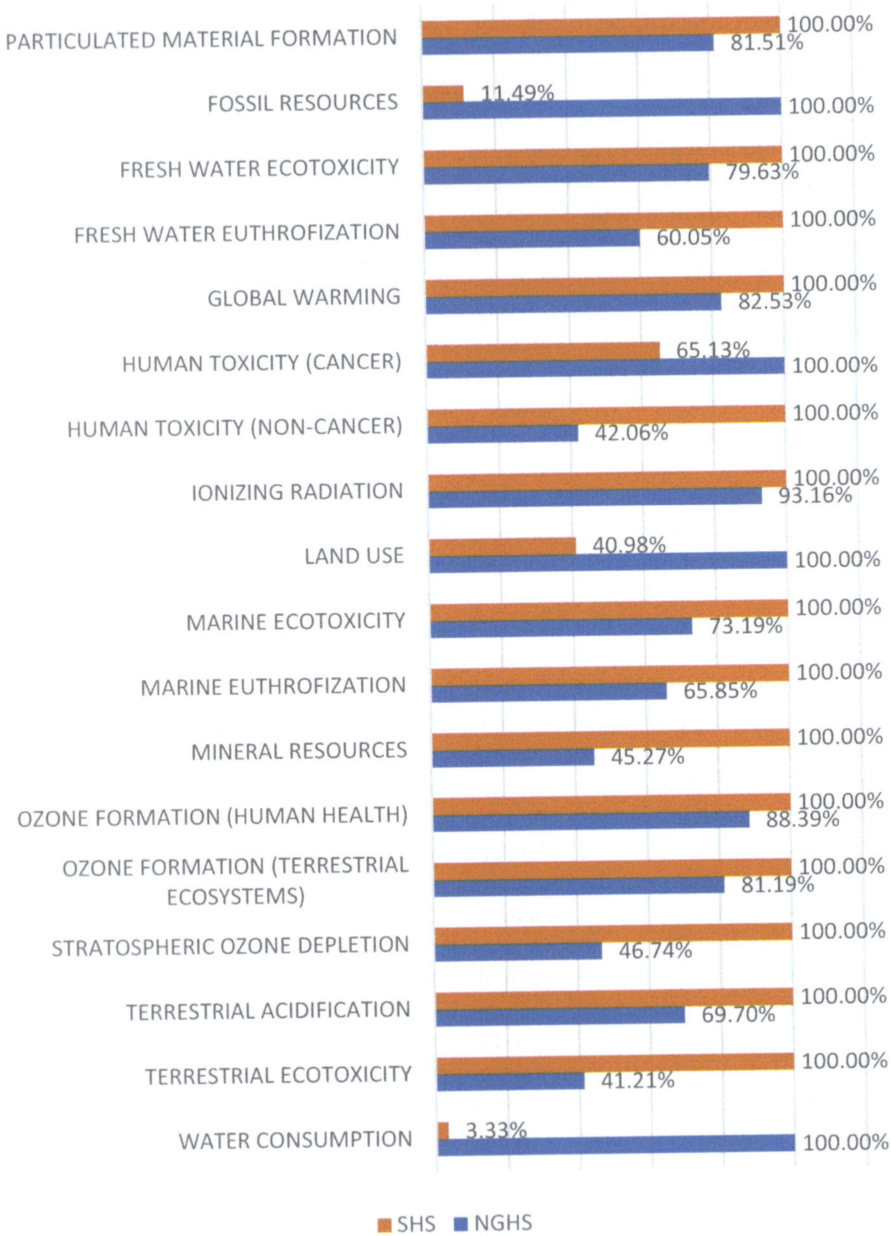

Fig. 4 Normalized results for the ReCiPe midpoint impact categories (H)

the operational phase, considering that, according to the Fig. 4, approximately (73%) of the impacts generated are associated with the supply of natural gas to the system.

In conclusion to the results obtained, it can be inferred that the HWBS with heating via SHS has better environmental performance than the system with heating via Natural Gas, even though the first one uses an electrical complement for operation appropriate. In most of the impact categories analyzed, the system in question is more advantageous than gas, a factor that must be associated mainly with the consumption of energy resources throughout its useful life, considering that the use of irradiation solar contributes a lot to the decrease in energy demand by another source, even if it does not eliminate it entirely. Furthermore, another important factor for performance was the use of CPVC piping for distribution. In the case of the gas system, the use of copper piping has a strong influence on the addition of several analyzed impacts. Finally, it is important to highlight that the SHS has a higher volume of infrastructure applied to the assembly process, causing higher initial resource consumption, however, this factor is balanced, over time, by lower consumption of energy resources for operation.

5 Conclusions

This work aims to present a proposal for a method derived from the general Life Cycle Assessment methodology in order to compare the environmental performance of two distinct Hot Water Building System for multi-family residential developments, through thermal heaters installed on the final roof of buildings, with supplementation of electrical supply, so that accurate information on the environmental performance of the systems can be obtained. However, the installed distinct Hot Water Building System considered in this work are Natural Gas Heating System and Solar Heating System. The novelty of this work is in the application of an environmental management method to empower the decision-making process and encourage the selection course of Hot Water Building System, considering the technical and economic aspects at an early designing phase of buildings.

The proposed method recommends the analysis of the life cycle of Hot Water Building System still in the initial phases of building design, from a cradle-to-gate perspective, including primary extraction of the component materials, processing and production and transportation to the place of execution of the projected installation, so that the energy incorporated into them and their biogenic emissions are considered and give greater dimension to the impacts of the systems, as well as considering information on the use profile of these building systems over its useful life, such as water and energy consumption for heating, bringing to the beginning of the lifecycle of buildings a range of information that allows the designer to make a clearer decision regarding which component system and materials to use in the project and its reflexes to the building and its surroundings.

The proposed method of this work has some limitations regarding its implementation. For example, the results of the application will provide a complex set of numerical values for the environmental impact indicators and a report with all related assumptions elaborated during the analysis, which makes interpretation of results by non-specialists in Life Cycle Assessment difficult, especially if there is no comparative pattern, which is a notable determining factor in the results of the analysis carried out for buildings and building systems, which they still relate to external triggers such as customer demands, government incentives or regulatory obligations acting on the context of formation and installation of building projects. Another limitation refers to the lack of an inventory database for South America, and, for local studies, it is usual to apply Life Cycle Assessment using existing data on global bases, a factor that distances the results from reality application locations. Still on the regionality of the studies, it is necessary to highlight the heterogeneity of the natural, cultural and economic profiles among the localities of the region, a factor that distances even more results based on global databases from a localized reality.

On the other hand, the determination of a methodological standardization regarding the application of Life Cycle Assessment for building systems allows more professionals to be involved in this process, and data can be obtained from the results of applications in buildings, which can be a parameter for other studies on similar buildings and promote an information chain that can bring. The objective of this work is to obtain precise information about the environmental performance of Hot Water Building System as an alternative to the conventional system widely used in Brazil, which consists of heating by electric means, concludes that the System via SHS is the one with the best environmental performance when compared to the system via NGHS. However, as reported, many variants must be taken into account with regard to the data obtained and their reliability.

The size of the Hot Water Building System has a direct influence on the potential environmental impacts, since SHS needs a much more robust installation infrastructure than the heating system via Natural Gas, a factor that can weigh on its performance. The gas system uses a fossil heating source, which has a potentially polluting production chain, while the solar system uses natural radiation available in the atmosphere. In this way, the time of use can counterbalance the installation factor with respect to the potential polluter: an analysis that encompasses the entire useful life tends to favor the application of the solar system to the detriment of gas. Factors of real importance for the data obtained still consist of the regionalization of the application and the uniqueness of the systems, considering that the architectural layout of the buildings is mandatory for the arrangement of the facilities, the place of application and the availability of resources.

The collected results may have significant variability of some magnitude being altered, and may not accurately reflect the local reality, also considering the use of international databases, such as Ecoinvent, to obtain information about the production chain of the building materials for building systems. However, the important value obtained, which consists in guiding the profile of the implementation of the hot water building system for a residential development, is maintained as, with the results, one can have an idea of which system is more advantageous to implement for undertakings of the same size as the one under study, also considering factors external to the environmental issue, such as the availability of installation and maintenance resources throughout the operational life of the installed equipment. Hence, it is suggested that data improvement be carried out with regard to obtain regionalized data for the national market, and even more for regional markets within the scope of the projects, so that a database can be created as a result of applying similar studies to building systems for use by designers and companies in the design decisions of enterprises and their subsystems.

References

1. W.B. Meyer, B.L. Turner, Human population growth and global land-use/cover change. Annu. Rev. Ecol. Syst. **23**(1), 39–61 (1992)
2. J.P. Harte, Human population as a dynamic factor in environmental degradation. Popul. Environ. **28**(4–5), 223–226 (2007)
3. K. Riahi, S. Rao, V. Krey, C. Cho, V. Chirkov, G. Fischer, G. Kindermann, N. Nakicenovic, P. Rafaj, RCP 8.5—a scenario of comparatively high greenhouse gas emissions. Clim. Change **109**, 33 (2011)
4. N. Fumo, P. Mago, R. Luck, Methodology to estimate building energy consumption using EnergyPlus Becnhmark models. Energ. Build. **42**(1), 2331–2337 (2010)
5. A.B. Constantinos, K. Droutsa, E. Dascalaki, S. Kontoyiannidis, Heating energy consumption and resulting environmental impact of European apartment buildings. Energ. Build. **37**(5), 429–442 (2005). https://doi.org/10.1016/j.enbuild.2004.08.003
6. A.D. Valdehi, R.V. Ralegaonkar, S. Mandavgane, Improving environmental performance of building through increased energy efficiency: a review. Sustain. Cities Soc. **1**(4), 211–218 (2011). https://doi.org/10.1016/j.scs.2011.07.007
7. G. Martinopoulos, K.T. Papakostasa, M. Papadopoulos, A comparative review of heating systems in EU countries, based on efficiency and fuel cost. Renew. Sustain. **90**(1), 687–699 (2018). https://doi.org/10.1016/j.rser.2018.03.060
8. M. Najjar, K. Figueiredo, A.W.A. Hammad, A. Haddad, Integrated optimization with building information modeling and life cycle assessment for generating energy efficient buildings. Appl. Energ. **250**, 1366–1382 (2019)
9. S. Glick, A.A. Guggemos, Life-cycle assessment and life-cycle cost as collaborative tools in residential heating system selection. J. Green Build. **5**(3), 107–115 (2010)
10. H.B. Randi, A.E. Marc, A review of the sustainability of residential hot water infrastructure: public health, environmental impacts, and consumer drivers. J. Green Build. Fall **6**(4), 77–95 (2011)
11. C.C.S. Moore, E.E. Rego, L. Kulay, The Brazilian electricity supply for 2030: a projection based on economic, environmental and technical criteria. Environ. Nat. Resour. Res. **7**(4), 17–29 (2017). https://doi.org/10.5539/enrr.v7n4p17
12. W. Klöpffer, Introducing life cycle assessment and its presentation in 'LCA compendium, in W. Klöpffer (eds.), *Background and Future Prospects in Life Cycle Assessment. LCA Compendium—The Complete World of Life Cycle Assessment* (Springer, Dordrecht, 2014), pp. 39–84
13. UNEP, *Avaliação de Políticas Públicas para Redução da Emissão de Gases de Efeito Estufa em Edificações* ([s.n.], São Paulo, 2012). Disponível em http://www.cbcs.org.br/userfiles/comitestematicos/outrosemsustentabilidade/UNEP_capa-miolo-rev.pdf
14. C.G. De Souza, R.G. Barbastefano, R.C. Teixeira, Life cycle assessment research in Brazil: characteristics, interdiciplinarity, and applications. Int. J. Life Cycle Assess. **22**(1), 266–276 (2017). ISSN 0948-3349

15. O. Coelho Filho, N.L. Saccaro Jr., G. Luedemann, *A avaliação de ciclo d vida cmo ferramenta para a formulação de políticas públicas no Brasil* (Ipea, Brasília, 2016)
16. E. Meex, A. Hollberg, E. Knapen, L. Hildebrand, G. Verbeeck, Requirements for applying LCA-based environmental impact assessment tools in the early stages of building design. Build. Environ. **133**(1), 228–236 (2018)
17. J. Ochsendorf et al., *Methods, Impacts, and Opportunities in the Concrete Building Life Cycle* ([s.n.], Cambridge, MA, 2011). Disponível em https://www.greenconcrete.info/downloads/MITBuildingsLCAreport.pdf
18. G.A. Da Silva et al., *Avaliação do ciclo de vida: ontologia terminológica* (Instituto Brasileiro de Informação em Ciência e Tecnologia—Ibict [s.l.]: [s.n.], 2014), p. 72
19. ISO. International Organization for Standardization, *ISO 14040. Environmental Management—Life Cycle Assessment—Principles and Framework* (ISO, Geneva, 2006a), p. 20
20. ISO. International Organization for Standardization, *ISO 14044. Environmental Management—Life Cycle Assessment—Requirements and Guidelines* (ISO, Geneva, 2006b), p. 46
21. H.K. Stranddorf, L. Hoffmann, *Update on Impact Categories, Normalization and Weighting in LCA-Selected EDIP97-data*, vol. 995 (Danish Environmental Protection Agency, 2005a), pp. 290 [s.l.]
22. M. Najjar et al., Integration of BIM and LCA: evaluating the environmental impacts of building materials at an early stage of designing a typical office building. J. Build. Eng. **14**(1), 115–126 (2017)
23. M.K. Najjar, K. Figueiredo, A.C.J. Evangelista, A.W.A. Hammad, V.W.Y. Tam, A. Haddad, Life cycle assessment methodology integrated with BIM as a decision-making tool at early-stages of building design. Int. J. Constr. Manag. 1–15 (2019b)
24. N.M. Crespo, C. Bueno, A.R. Ometto, Avaliação de Impacto do Ciclo de Vida: revisão dos principais métodos Palavras-chave. Prod. (x) (2013)
25. G. Machado, *Aprenda como funciona a gestão de Stakeholders na Construção Civil.* (Halo Notoriedade Empresarial, 2017). Disponível em http://halonotoriedade.com.br/aprenda-como-funciona-a-gestao-de-stakeholders-na-construcao-civil/. Acesso em 07/abr./19
26. J. Oyarzo, B. Peuportier, Life cycle assessment model applied to housing in Chile. J. Clean. Prod. **69**(1), 109–116 (2014). ISSN 0959-6526
27. A. Inaba et al., Chapter 4: Data documentation, review, and management, in *Global Guidance Principles for Life Cycle Assessment Databases: A Basis for Greener Processes and Products* ([s.l.], 2011), pp. 85–95. ISBN 978-92-807-3174-3
28. J.C. Bare, P. Hofstetter, D.W. Pennington, H. Udo de Haes, Int. J. LCA **5**(319) (2000)
29. A.P. Acero, C. Rodríguez, A. Ciroth, *LCIA Methods: Impact assessment methods in Life Cycle Assessment and their impact categories* ([s.l.]: [s.n.], Greendelta, 2015), p. 23
30. RIVM, *LCIA: The ReCiPe Model* (2018). Disponível em https://www.rivm.nl/en/life-cycle-assessment-lca/recip

The manufacturer's authorised representative in the EU is Springer Nature Customer Service Centre GmbH, Europaplatz 3, 69115 Heidelberg, Germany. If you have any concerns regarding our products, please contact ProductSafety@springernature.com

Printed and bound by CPI Group (UK) Ltd, Croydon, CR0 4YY
22/09/2026
02220948-0019